RETHINKING
LOS ANGELES

RETHINKING LOS ANGELES

EDITED BY

MICHAEL J. DEAR
H. ERIC SCHOCKMAN
GREG HISE

PUBLISHED IN ASSOCIATION WITH THE
SOUTHERN CALIFORNIA STUDIES CENTER
OF THE UNIVERSITY OF SOUTHERN CALIFORNIA

SAGE Publications
International Educational and Professional Publisher
Thousand Oaks   London   New Delhi

For information address:

SAGE Publications, Inc.
2455 Teller Road
Thousand Oaks, California 91320
E-mail: order@sagepub.com

SAGE Publications Ltd.
6 Bonhill Street
London EC2A 4PU
United Kingdom

SAGE Publications India Pvt. Ltd.
M-32 Market
Greater Kailash I
New Delhi 110 048 India

Printed in the United States of America

Library of Congress Cataloging-in-Publication Data

Main entry under title:

Rethinking Los Angeles / editors, Michael J. Dear, H. Eric Schockman,
    Greg Hise.
        p.   cm.
    Includes bibliographical references and index.
    ISBN 0-8039-7286-5 (cloth: acid-free paper).—ISBN
0-8039-7287-3 (pbk.: acid-free paper)
        1. Los Angeles Metropolitan Area (Calif.)—Social conditions.
    2. Los Angeles Metropolitan Area (Calif.)—Politics and government.
    3. Urban policy—California—Los Angeles Metropolitan Area.
    I. Dear, M. J. (Michael J.)   II. Schockman, H. Eric.   III. Hise,
    Greg.   IV. Series.
    HN80.L7R47 1996
    306'.0979'494—dc20   96-10037

This book is printed on acid-free paper.

96 97 98 99 10 9 8 7 6 5 4 3 2 1

Sage Production Editor: Diana E. Axelsen
Sage Typesetter: Andrea Swanson
Cover Photo: Robbert Flick

Contents

Preface

The Los Angeles region is increasingly held up as a prototype (for good or ill) of our collective urban future. Yet it is probably the least understood, most understudied major city in the United States of America. Few persons beyond the boundaries of Southern California have an accurate appreciation of what the region is, who lives there, and what it does. Instead, popular perceptions rely on the exaggerated images promulgated by television and movies and by print media predictions of the impending Southern California apocalypse. This book brings together some important voices to dispel the myths about Southern California and to begin the process of "rethinking" Los Angeles.

The beautifully named *El Pueblo de Nuestra Señora la Reina de los Ángeles de Porciúncula* was founded in 1781. One hundred and fifty years later, a pamphlet issued by the Security Trust and Savings Bank (Hill, 1929) immodestly proclaimed, "Thus came into being *a* great American city, destined, as many predict, to be *the* great American city" (p. 12). Boosterism has persisted as an integral part of L.A.'s image and purpose throughout its history. In a more recent plan for the year 2000, local visionaries (Los Angeles 2000 Committee, 1988) imagined the imminent emergence of a "world cross-roads city," in which, just as in times past, citizens would be able to make their "Los Angeles dream come true" (p. 86). Of course, there has always been a dark side, too. Bertolt Brecht (1976), fleeing the Third Reich, found "on thinking about Hell, that it must be/Still more like Los Angeles" (p. 367). The city's bleak urbanism has also provided the source of countless gaudy and grotesque parodies in literature and film. Perhaps best known is Ridley Scott's *Blade Runner,* which evoked a futuristic metropolis of perpetual darkness, a polyglot inferno of high-tech crime. But Roman Polanski's *Chinatown* captures the reality more accurately: a superficial gloss of striking beauty, glowing light, and pastel hues, which together conspire to conceal a hideous culture of malice, mistrust, and mutiny.

Casual observers, visitors and residents alike, catch little of the city's underbelly. They are instead persuaded by the glossy, utopian images of the burgeoning World City—a collage of prosperity, fantasy, and play: the corporate glitter of a downtown citadel; the sunshine, surf, and mountains; the city as a giant agglomeration of theme parks. Beneath such images is a cityscape more reminiscent of a Third World nation, a dystopia that is increasingly polarized between haves and have-nots, in which neighborhoods increasingly resemble combat zones as warring gangs struggle for turf supremacy. Here, the air, earth, and water are perpetually being poisoned. Here, public responsibility for basic human services, including shelter, education, and health care, are being abdicated.

The origins of this book lie directly in the events of April and May of 1992, when some of the worst unrest in U.S. urban history transformed Los Angeles into a fire zone. The most immediate precursors of this upheaval were the verdicts associated with the trial of the police officers involved in the beating of Rodney King. Most people, however, now concede that the civil disturbances were but one more episode in the continuing history of racial discrimination and inequality that have plagued our cities for centuries. The events following the verdicts have been called many things, including a riot, an economic referendum, a rebellion, civil unrest, and civil disturbance. They were, in truth, all of these things, but they can most readily be understood as an expression of outrage by peoples who find themselves increasingly marginalized in our changing society, without hope of realizing their vision of the American dream.

The events of 1992 prompted a continuing series of national, regional, and local conversations about the future of urban America. Throughout 1993 and 1994, the campus of the University of Southern California hosted many such dialogues. Among the most prominent was a public discussion series titled "Rethinking Los Angeles," which brought together campus scholars and community representatives to discuss ways of creating a better future for Southern California and its people. This book is a collection of the contributions to that conversation; it consists of a sequence of formal chapters interleaved with the comments of prominent members of the community of Southern Californians.[1] More than anything else, these chapters and commentaries indicate that no one group (be they scholars, activists, or community leaders) has a monopoly of wisdom about the city. On the contrary, these conversations convey just how important it is to bring people together to pool their collective experience about ways of re-creating the city.

There was one more impetus for this volume. During the 1980s, a group of loosely associated scholars, professionals, and advocates based in Southern California became self-consciously aware that what was happening in the Los Angeles region was somehow symptomatic of a broader transformation taking place in urban America. Just as the Chicago school emerged in the 1930s when that city was reaching new national prominence as the prototypical industrial metropolis, Los Angeles is already imprinted on the minds of a new generation of international urban scholars. The fundamental contrast between Chicago and Los Angeles is this: Although Chicago was characterized as a monocentric city that spread outward from a dense core in rings of ever decreasing density and relatively orderly land use transitions, Los Angeles was from its inception always a low-density, polycentric urban agglomeration in which development occurred in a patchwork quilt of mixed uses. The polycentrism of Southern California has become more exaggerated with the passage of time, compounded today by the burgeoning multicultural and polyglot character of the populations in the sprawling metropolis. The present-day confluence of global changes in economic organization, political structures, and sociocultural practices that Southern California is experiencing almost *requires* the development of alternative theories of urban growth. Sometimes, these theories overlap and coexist in their explanations of the emerging metropolis. The proliferation of diverse ways of understanding is consistent with the project of postmodernism; it is no coincidence that Los Angeles is held by some to be the prototypical postmodern city.

Southern California is undoubtedly a special place. But adherents of the "Los Angeles school" do not argue that the city is unique or that it is necessarily a harbinger of the future, although both viewpoints are at some level demonstrably true. It is simply that at present, an especially powerful intersection of empirical and theoretical research projects has come together in a particular place at a particular time, that these projects have attracted the attention of a critical mass of scholars and practitioners, and that the world is facing the prospect of a Pacific century in which Southern California is likely to become a global capital. The strength and potential of the Los Angeles school derive principally from the intersection of these events and the promise they hold for the re-creation of urban theory.

The chapters and commentaries in this collection focus on Southern California in all its diversity. We are firmly convinced that the region can be understood only by reexamining all aspects of its livelihood, peoples, and environment. The chapters that follow examine aspects of the regional economy, politics, society, and culture. We dig deeply into its history and geography and address important problems concerned with art, health, environment, education, policing, immigration, and race and ethnicity. One word of advice for our readers: There is no recommended way to read this book. Although most people will want to start at the beginning, it is also possible to selectively sample from the collection of chapters and commentaries at any point. The chapters have been informally arranged around several key themes:

- the problems involved in understanding what we observe in the landscapes of Southern California (Chapters 1 and 2);
- the past and future Southern Californian economy (Chapters 3 and 4);
- the politics of reform in the governments of the city and county of Los Angeles (Chapters 5 and 6);
- dealing with pivotal questions of race and ethnicity (Chapters 7, 8, and 9);
- the environment (Chapter 10);
- the issues of public (and private) services, including education, health, and transportation (Chapters 11, 12, and 13); and
- envisioning the urban future (Chapter 14).

Another important feature of all the chapters is that they are not simply concerned to describe the situation as it exists. Instead, the authors are committed to re-imagining alternative futures for Southern California. In this endeavor, they do not speak with a single voice; we should not expect uniformity of opinion. Several common—sometimes contradictory—themes, however, have emerged from these diverse contributions, in a totally unorchestrated fashion. These are *a sense of lost community, or even lost soul, in the City of the Angels,* which must be set alongside an *intense commitment to rewriting the social contract that brings people together* in Southern California. There is also an *aura of regret or sadness, almost an elegaic quality* to many of the chapters and commentaries, as our contributors look back on the events of the recent past; but once again, this is *offset by the pervasive sense of optimism that things can be made better, that the reconstructive effort is already under way.*

As editors, we cannot speak for our contributors. But one thing is clear after our collective effort to rethink Los Angeles. The city can no longer be regarded as the exception to the patterns of American urbanism. We are convinced that the urban agenda facing Los Angeles has much wider implications. The social contract that brings people together in Southern California is currently being renegotiated; there is reason to believe that if this renegotiation fails in Los Angeles, then it can succeed nowhere else in this country.

Note

1. Comments of these community members are used with permission.

References

Brecht, B. (1976). On thinking about hell. In B. Brecht, *Poems 1913-1956* (p. 367). London: Part Three.
Hill, L. L. (1929). *La Reina: Los Angeles in three centuries* [Pamphlet]. Los Angeles: Security Trust and Savings Bank.
Los Angeles 2000 Committee. (1988). *Los Angeles 2000: A city for the future.* Los Angeles: Author.

Acknowledgments

In preparing this volume, we have received a great deal of assistance. We would especially like to thank our contributors for providing excellent materials and responding to our many requests in a timely fashion.

The original yearlong public lecture series on which this book is based was funded by several units on the campus of the University of Southern California, including the College of Letters, Arts, and Sciences; the Center for Multiethnic and Transnational Studies; the Center for Research on Environmental Science, Policy, and Engineering; the Department of Geography; the Division of Social Sciences and Communications; the School of Education; the School of Public Administration, and the School of Urban and Regional Planning.

Funding was also provided by the Getty Center for the History of Art and the Humanities, and we wish especially to thank Tom Reese and Lynn O'Leary Archer for their support, as well as Jo-Ellen Williamson for administrative assistance.

We are deeply grateful for the collegial support and advice received from Allen Scott and Roger Waldinger, respectively, past and present directors of the Lewis Center for Regional Policy Studies at the University of California at Los Angeles. Thanks also to Mark Ellis and to Jennifer Wolch, both of whom made valuable comments on the entire text and much improved the final product.

Carrie Mullen, our editor, departed from Sage Publications just as this volume was completed. She was a great champion of this book, and we gratefully acknowledge her efforts on our behalf.

The excellent production skills of Clare Walker aided in the preparation of the manuscript. Frank Sommers acted as research assistant and played an important role in the implementation of the Rethinking Los Angeles public lecture series. Michael Preston, Alvin Rudisill, and C. S. Whitaker offered valuable intellectual guidance and support. Steven H. Crithfield served as a partner in life.

Work on this volume was completed when Dear was a fellow at the Center for Advanced Study in the Behavioral Sciences at Stanford. The support of the center and the National Science Foundation (#SES-9022192) is gratefully acknowledged.

Publication of this volume was made possible by a grant from the Southern California Studies Center of the University of Southern California.

We have reserved until last special thanks to the crew at Sage Publications who were responsible for bringing the book into production so efficiently, and who (incredibly!) turned the tasks of copyediting, proofreading, etc., into enormous fun. Our thanks go to Diana Axelsen (Production Editor), Ravi Balasuriya, Alison Binder, Jessica Crawford, Jennifer Morgan, Renée Piernot, Catherine Rossbach, Andrea Swanson, and Dianne Woo.

Folio: On Pico Boulevard Looking North (December, 1994)

ROBBERT FLICK

Professor of Fine Arts,
University of Southern California

Folio: On Pico Boulevard Looking North (December, 1994)

ROBBERT FLICK
Professor of Fine Arts, University of Southern California

1. From Appian Way to 4th Street, Santa Monica
   (000 Pico to 400 Pico)

2. From Yorkshire Avenue to Urban Avenue, Santa Monica
   (2900 Pico to 3300 Pico)

3. From Camden Avenue to Westwood Boulevard, West Los Angeles
   (11000 W. Pico to 10800 W. Pico)

4. From Redondo Boulevard to Longwood Avenue, Los Angeles
   (5000 W. Pico to 4800 W. Pico)

5. From San Vicente Boulevard to Lucerne Boulevard, Los Angeles
   (4600 W. Pico to 4400 W. Pico)

6. From Norton Avenue to 4th Avenue, Los Angeles
   (4000 W. Pico to 3800 W. Pico)

7. From Vermont Avenue to Magnolia Street, Los Angeles
   (2400 W. Pico to 2200 W. Pico)

8. From Main Street to Santee Street, Los Angeles
   (100 E. Pico to 300 E. Pico)

Location of Cover Photo Segments

2600 block E. Whittier Boulevard at Mott Avenue, East Los Angeles (looking north)

300 block N. Rodeo Drive, Beverly Hills (looking east)

900 block S. Vermont Avenue, Korea Town (looking west)

10800 block W. Pico Boulevard, West Los Angeles (looking north)

7600 block Melrose Avenue at Stanley, Hollywood (looking north)

200 block E. Pico Boulevard, Los Angeles (looking north)

900 block N. Broadway, Chinatown (looking north)

500 block S. Broadway, downtown Los Angeles (looking south)

700 block S. Alvarado Street, Westlake (looking east)

2000 block W. Sunset Boulevard at Alvarado, Echo Park (looking south)

For quicktime video of these segments, see
http://www.usc.edu/dept/matrix/

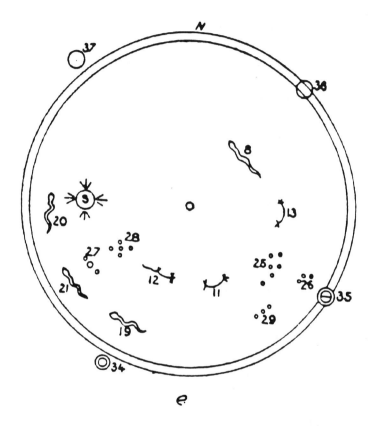

Figure 1.1. The first known "map" of the Los Angeles region, in the form of a Diegueño Indian ground, or sand, painting. The universe is oriented with north at the top. The circle numbered 37 in the upper left represents Catalina Island, southwest of Los Angeles. The circle numbered 36 in the upper right is San Bernardino Mountain. Circles on the southern horizon represent landmarks of the San Diego area. Figures within the circle usually represent astronomy or wildlife, with snakes being prominent on the southern margins.

SOURCE: Kroeber, 1925, Figure 56e.

Rethinking Los Angeles

GREG HISE

MICHAEL J. DEAR

H. ERIC SCHOCKMAN

Every single American city that *is* growing, is growing in the fashion of Los Angeles.

Joel Garreau, 1991, p. 3

Los Angeles has almost always been viewed as an exception to the mainstream of American urban culture. Located on the continental margins, the city conjures up visions of unbridled urban sprawl, inconsequential architecture, freeways, sun, surf, and smog. Such images have been exaggerated by the movies and television that Hollywood has sold to the world. In many ways, Los Angeles *is* unlike other cities. For instance, it is strikingly different from Chicago, the city that has been universally regarded as the prototypical industrial metropolis. For many decades, urbanists have analyzed the cities of the world according to precepts of the Chicago school. There is, however, a deep problem with the exceptionalist narratives of Los Angeles. They render much of what happens in the city as merely illustrative, a series of quirky set pieces, even staged performances. Yet at least in appearance, Los Angeles bears a strong resemblance to other emerging "world cities," such as Mexico City and São Paulo, as well as fast-growing urban centers in the United States, including Atlanta, Phoenix, and Seattle.

Fortunately, during the past decade, increasing scholarly attention has been directed toward the city of Los Angeles and the five-county region that composes Southern California (Los Angeles, Ventura, San Bernardino, Riverside, and Orange Counties).[1] Drawing on this evidence, scholars and pundits in the emergent "Los Angeles school" have formulated new theories that challenge the prevailing assumptions concerning the urban future. The Chicago school model with its fixed core and dependent rings was cast at a time when U.S. cities were decanting at an unprecedented rate. The Southern California prototype, with its emphasis on multicentered, dispersed patterns of relatively

1

low-density growth and multicultural/multiethnic enclaves, may yet become a new paradigm of metropolitan development—what some people refer to as a postmodern urbanism.

In population, Los Angeles is now the second largest metropolitan region in the United States and 11th largest in the world. Forecasters expect it will have more than 20 million people by the year 2000. The five-county region, encompassing approximately a 60-mile circle centered on downtown Los Angeles, contains only 5% of California's total land area. Yet within this circle resides more than half the state's population and personal income. Fifty-six percent of the state's international trade and the headquarters for 58 of the 100 largest companies are located here. The gross product per person in the 60-mile circle ranks fourth in the world. The region is also known for its extraordinary consumption patterns. In Los Angeles alone, more than 2,000 cars are sold daily, including 20% of all U.S. Rolls Royce registrations and 70% of all California registrations. At the same time, however, Los Angeles is known as the "homelessness capital" of the United States, and one in seven of the county's residents relies on some form of public assistance.

Understanding Los Angeles

The essence of Los Angeles was revealed more clearly in its deviations from [rather] than its similarities to the great American metropolis of the late nineteenth and early twentieth centuries. (Fogelson, 1967, p. 134)

How do we begin to understand Los Angeles and Southern California? How can we explain what is happening in the region? Most world cities have an instantly identifiable signature: Think of the boulevards of Paris, the skyscrapers of New York, or the churches of Rome. But Los Angeles appears to be a city without a common narrative; nomenclatures such as "Hollywood" or "Beverly Hills" are universally understood as fragmentary distortions of some broader, more opaque canvas. Twenty years ago, Reyner Banham (1973) provided an enduring map of regional landscapes. To this day, it remains powerful, evocative, and instantly recognizable. Banham identified four basic "ecologies": surfurbia (the beach cities); the foothills (the privileged enclaves of Beverly Hills, Bel Air, etc., for which the financial and topographical contours correspond almost exactly); the plains of Id (the endless central flatlands; chap. 8); and autopia (the freeways, a "complete way of life" to Banham; p. 213).

For Douglas Suisman (1989), the boulevards, not the freeways, determine the city's overall physical structure. A boulevard is a surface street that makes arterial connections on a metropolitan scale, provides a framework for civic and commercial destinations, and acts as a filter to adjacent residential neighborhoods. Suisman argues that boulevards do more than establish an organizational pattern; they constitute "the irreducible armature of the city's public space" (p. 7) and are, therefore, charged with social and political significance. Today, these connectors form an integral link among the region's municipalities and provide a fundamental key to unlocking the local cartography. John Gregory

Dunne once pointed out that to understand Sunset Boulevard is to be on the way to understanding the meaning of Los Angeles.

For Edward Soja (1989), Los Angeles is a decentered, decentralized metropolis powered by the insistent fragmentation of "post-Fordism," an increasingly flexible, disorganized regime of capitalist accumulation. Accompanying this shift is a postmodern consciousness, a cultural and ideological reconfiguration altering how people experience social being. The government center in downtown Los Angeles functions as an urban panopticon—a strategic surveillance point for the state's exercise of social control. From the center extends a patchwork quilt of "wedges" and "citadels," urban islands formed by the boulevard corridors. The consequent urban structure is complex and fragmented yet bound to an underlying economic rationality. Soja concludes, "With exquisite irony, contemporary Los Angeles has come to resemble more than ever before a gigantic agglomeration of theme parks, a lifespace composed of Disneyworlds" (p. 246).

These sketches provide differing insights into the landscapes of Southern California. Banham considers the city's overall torso, recognizing three basic elements (suburbia, plains, foothills) plus the connecting arteries (freeways). Suisman shifts our gaze from the principal arteries to the veins that channel everyday life (the boulevards). Soja considers the body-in-context, articulating the links between political economy and postmodern culture to explain polarization and fragmentation in the city.

In this chapter, we focus primarily on some of the myths of diversity and complexity of life in Los Angeles. We first consider the "past realities" that have and continue to inform perceptions about Southern California, especially the proliferation of evocative myths about the region and its dreamscapes. Then we turn our attention to "present realities" and begin the enormous task of *rethinking* contemporary Los Angeles. Along the way, we also introduce the diverse contributions to this volume.

Realities Past: Forty Myths in Search of a City

> Penthesilea is different. You advance for hours and it is not clear . . . whether you are already in the city's midst or still outside. Like a lake with low shores lost in swamps, Penthesilea spreads for miles around. . . . Every now and then at the edges of the street a cluster of constructions . . . seems to indicate that from there the city's texture will thicken. But you find instead other vague spaces. . . . And so you continue, passing from outskirts to outskirts [until] you have given up trying to understand whether [there] exists a Penthesilea the visitor can recognize and remember, or whether Penthesilea is only the outskirts of itself. (Calvino, 1974, pp. 156-157)

Like modern-day Marco Polos encountering one of Italo Calvino's "continuous cities," visitors, residents, and commentators alike are struck by the seemingly endless expanse of the Los Angeles metropolitan area. Invariably these accounts employ familiar terms such as *suburbs* and *sprawl* with *automobility, freeways,* and *single-family housing* as standard links in a causal chain of explanation. More often than not, a palpable concern is evident in these discussions. Urban expansion, both physical and social, is seen as an uncontainable, possibly mutant force; regional growth not only is unplanned and without

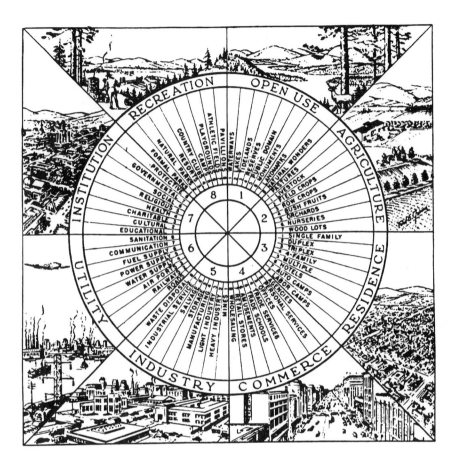

Figure 1.2. The yearning for rationality in the development of the Southern California urban landscape is expressed in this logo, which decorated many of the land use planning documents of the 1920s.

order but also is simply beyond control. Although this process, typically described as *suburbanization,* is not limited to Los Angeles or Southern California, the region is perceived as the prototype of urban fragmentation—the place where, contrary to the city's official motto, it all comes apart.

As part of our rethinking process, we want to consider why outsiders clung so tenaciously to these and other myths as the "truth" about Southern California. To cite but one example: In 1962, a visiting British academic named Richard Gilbert (1964) decided to try his hand at pinning down the meaning of Los Angeles. At the end of the year, he discovered precisely what others have found—that the automobile set the city's basic pattern, that freeways sucked the life out of downtown, and that eventually urban Europe (and perhaps the rest of the world) would come to look like Los Angeles.

Fragmentation is a recurrent theme. In 1928, Sarah Comstock, a journalist writing for *Harper's Magazine,* described the city in astronomical terms, as a constellation of planets frozen in orbit.

> Go to Mount Wilson at night and you will be told that you are looking at Los Angeles and her satellites of sixty towns. They might as well tell you six hundred. Nothing like it is anywhere to be seen upon the earth's surface. Constellations such as you never dreamed of blaze in that valley below; thick star clusters, dripping Milky Ways, with here and there fat moons reddening. A woven pattern of lights, vast and intricate beyond description. (p. 721)

Three years prior to Comstock's observation, movie mogul and erstwhile urban theorist Cecil B. DeMille (1925) penned a gastronomic metaphor to describe L.A.'s urban pattern. He found that through providence and "good planning," residential and industrial districts as well as mixed satellites such as Colegrove and Garvanza had been scattered in and about Los Angeles like "plums in a fruit cake" (p. 49). DeMille anticipated, or at least encouraged, the region's continued expansion along these lines and thanked local industrialists for dispersing their enterprises, thus preparing the region for "enviable growth" (p. 49).

How can we account for these divergent perspectives—on the one hand, anarchic sprawl and on the other, increased opportunities through dispersion? Is it simply a discrepancy between a misinformed outsider and a booster's wishful fantasy? We think not. Language is important; rethinking Los Angeles requires us to examine the vocabulary and metaphors, both social and spatial, used in describing and interpreting cities. We might begin, for example, by vetoing use of the term *suburban sprawl* because it is so freighted with mythology and obsolete intellectual baggage; such terminology effectively obscures the nature of Los Angeles as well as of other metropolitan regions.

Language, myth, and landscape are all entwined in the narratives of the West and of Los Angeles in particular. Recognizing and interpreting contemporary sociospatial patterns require that we first uncover the intentions and perceptions of the historical agents of urban change. Los Angeles is often considered disparagingly as *anti-urban,* particularly when compared with those "real" cities in Europe, their counterparts along the Atlantic seaboard, and the old west (including London, Paris, New York, Chicago, Philadelphia, and Pittsburgh). Reformers, civic elites, and city builders in turn-of-the-century Los Angeles viewed things differently. For example, Dana Bartlett, a progressive proponent of good government, improved sanitation, and social harmony, saw the city becoming another New York or Chicago. Instead of acquiescing to this particular form of urbanism, he argued for the creation of a new type of city. Bartlett believed Los Angeles could be the first City Beautiful. In a 1907 reform tract titled *The Better City,* he wrote lyrically of the physical setting of Southern California, saving his grandest praise and prose for the Los Angeles River, which, at that time, was already crowded with "factories and the river bed itself is sought by utilitarian corporations." More important, Bartlett recognized that the river served as a physical and social divide. Bartlett foresaw the problems of spatial segregation and envisioned the river as a centerpiece for city renewal, redesigned as a tangible and symbolic link between the original town site on the east and wealthier Angelenos who had established new residential districts to the west.

Bartlett was not alone in his conscious attempt to fashion a different city. His was but one voice in a neglected, contrarian discourse that suggests Angelenos were not categorically anti-urban. Writing in 1910 for *Out West* magazine, William E. Smythe naturalized the creation of a new type of city in Los Angeles, one in which the "hovel [is] forestall[ed] by the bungalow and the tenement by the Garden City" (p. 297). Later, in 1941, Clarence Dykstra, the Department of Water and Power's "Efficiency Expert," urged residents to build a "cit[y] of glory and grandeur . . . a balanced structure, full of opportunities for work and play; [with] industry to provide for the needs of all" (p. 9). At the beginning of the postwar era, the indomitable and ever prescient Carey McWilliams (1949) discovered the "first modern, widely decentralized industrial city in America" (p. 28) in Southern California. This was not the product of happenstance. Smythe's terminology, Dykstra's evocation of balanced communities, and McWilliams's attention to decentralization undoubtedly bring to mind Ebenezer Howard, the British (later international) "garden city" movement, and those American planners and housing reformers identified with the Regional Planning Association of America (RPAA). Central to this lineage was the planned integration of land uses. Each dwelling would be in spatial proximity to schools, recreation, and shopping; every resident would have ready access to civic institutions, cultural facilities, and employment.

The proliferation of such plans and proposals are at odds with popular perceptions of Los Angeles. Most everyone considers it a "developers' city." Residents and critics have interpreted the city-building process here as unfettered and expansive growth and the region as undifferentiated, monotonous, and chaotic. Smythe (1910), for example, equated boundlessness with vitality and future promise, and the region's built environment has invoked both disorientation and exhilaration. Although Frankfurt school philosophers Horkheimer and Adorno portrayed a homogenized city as standardized as film and advertising, Tom Wolfe (1964) reveled in the "curvilinear forms of the freeways . . . and all the car fantasy architecture that go[es] with them" (p. 22). Wolfe also reminds us that the vision of entrepreneurs *and* government agencies has shaped Los Angeles. The received wisdom, however, tends to overlook the importance of planning ideals for those who built Los Angeles.

Significant documentation supports the claim that planners, builders, and residents believed Southern California could be the locale for a new urban configuration, which they strove to implement. In his article "Congestion de Luxe: Do We Want It?" Clarence Dykstra (1926) argued that contrary to the "centralization complex" manifest in East Coast cities, "the city of the future ought to be an harmoniously developed community of local centers and garden cities" (p. 395). Dykstra believed that Los Angeles held the promise of

> a great city population which for the most part lives near its work, has its individual [homes] and gardens, finds its market and commercialized recreational facilities right around the corner, and which because of these things, can develop a neighborhood with all that it means. (p. 398)

Contemporary graphic representations of plans, emblems, and official and popular icons offer insights into how boosters, pundits, planners, reformers, and citizens perceived and projected growth patterns in the region. In 1922, the Los Angeles County Board of Supervisors sponsored conferences on regional planning devoted to subdivision regulation, transportation, and water and sewage—issues whose effects and domain crossed over tidy jurisdictional boundaries. The proceedings from the first meeting, held in Pasadena, included an illustration depicting Los Angeles as an urban region consisting of a "center city surrounded by many satellite sub-center cities and communities" (Los Angeles County, 1922, p. 4). George Damon, an engineer and Regional Planning Commission board member, wrote the accompanying text: "The whole District is crystallizing about natural centers and subcenters. The nucleus is the business center of Los Angeles. Beyond the five- or six-mile circle we find sub-centers developing, each with its own individual character and identity" (p. 4). The accompanying diagram, abstracted from the regional system of automobile laterals, portrayed a set of four major and minor satellites. Glendale, Culver City, Montebello, and Palms appeared as minor subcenters in this metropolitan constellation. Conference representatives drafted a "Declaration of Inter-dependence," a pioneering regional planning document that catalogued their findings regarding specialization and units in an urban system. Although most accounts refer to dispersion in restrictive, residential neighborhoods, these pundits envisioned growth and development occurring in urbanized clusters within the metropolitan region. And, most significant, they based their projections on analyses of existing conditions in the region.

Conference participants fixed the region's gravitational center in L.A.'s "business center" or downtown. This formulation would not have surprised contemporaries. At that time (the 1920s), centralization and concentration were matters of primary public concern. During the preceding two decades, downtown Los Angeles had grown from a 12-block enclave centered on Main Street between Temple and Fourth streets to an area encompassing more than 60 square blocks. Angelenos, like their counterparts in medium to large cities throughout the country, were engaged in contested debates regarding land uses, activity densities, and transit. Photographs, newspaper accounts, fiction, and oral histories of the time reveal the mass of humanity and machines jostling for space on downtown streets. Whether it was department store shoppers crowding sidewalks along the Seventh Avenue retail district or businesspeople engaged in commerce on the Spring Street financial spine, the central business area was recognized as the metropolitan hub. People shared these streets, occasionally at their peril, with streetcars, automobiles, and an assortment of drayage vehicles. Contemporary observers considered this urban concentration a mixed blessing, understanding that it simultaneously was a product and symbol of commercial success and an impediment that might limit future expansion.

Two years after the regional planning conference, Angelenos voted to fund and implement a major traffic street plan. This document was drawn up by nationally known consultants, including Harland Bartholomew and Frederick Olmsted Jr., and set out a system of thoroughfares at 1-mile intervals (see Olmsted, Bartholomew, & Cheney, 1924). The subsequent street widenings and dedications produced the superblock pattern of boulevards and avenues that to this day continues to channel movement and inform the mental maps of residents and outsiders alike.

Thus, in a seemingly contradictory manner, Angelenos actively promoted the planned dispersion of people along with business, industry, and retail activities at the same time as they supported the development of a central business district equal to those in other major cities. Renderings from the period depict a future high-rise city with skyscrapers lining the boulevards, extending in some cases from downtown to the ocean. Even at this early stage in the region's urban history, dispersion was seen as an advantage, something to be planned. Business interests, real estate entrepreneurs, home builders, and home buyers recognized the significance of mass and private transit to this enterprise. Henry Huntington's Pacific Electric was a classic example of streetcar expansion.

Most accounts cast the railways and the automobiles that supplanted them as the primary agents of urban form in Southern California. In this view, the auto and federally funded highway projects opened up the urban fringe to leapfrogging residential subur-banization, even as the metropolitan region was the inevitable product of sprawl projected along transit routes. Fred W. Viehe (1981) and others have challenged the determinism implicit in the transportation-oriented accounts of regional growth. Viehe traced the development of "suburban industrial clusters" during the 1910s and 1920s, including the industrial suburb of El Segundo and its residential neighbor, Manhattan Beach. Viehe argued that L.A.'s dispersed morphology reflected the "desire to fulfill an industrial ideal based on manufacturing . . . rather than a rural ideal based on home and open spaces" (p. 17). In short, although a considerable transportation network connected existing settle-ments, it was not in itself a causal agent in their establishment.

The point then, is that we should jettison the conventional wisdom and tired lexicon that continue to shape our understanding of urban expansion and social relations in Southern California. Popular myths were fixed representationally in images showing the region as a composite of jurisdictions spreading across the basin, valleys, and contiguous counties—an undifferentiated sprawl. *Newsweek* portrayed this in a 1957 screed, "Suburbia—Exurbia—Urbia," which included a map of greater Los Angeles titled "Biggest Darn Collec-tion of Suburbs in the World," littered with the names of more than 90 districts and incorporated communities. This diagram reinforced a text that bemoaned the loss of cities as we knew them, in effect, a rehearsal of the sociologists' and design professionals' standard disdain for anything but a walking city with distinct districts and certain boundary. This critique was, in other words, nostalgic, normative, and prescriptive. On the ground, a casual perusal quickly dispels the notion that San Pedro and Lakewood, or Panorama City and Pasadena, were, or are, equivalent. They never were and never will be.

More important, they were not intended to be similar, nor were they conventional suburbs in the sense of economically inert bedroom communities populated by middle-and upper-income white families. Home builders and planners conceived of Westchester, Panorama City, and Lakewood as *self-contained* communities; they were not envisaged as an appendage to some urban core. Urbanists have to uncover these visions as an essential first step for constructing alternative interpretations of the Southern California experience. And contrary to popular conceptions regarding undifferentiated sprawl, Angelenos have and continue to parcel their turf into ever finer gradations. Gated enclaves are the least subtle form of this practice; it can also be found in the city council-sanctioned name plates littering streets, as well as in their spray-painted, criminalized counterparts.

Realities Present: Postmodern Los Angeles

But the city in its corruption refused to submit to the domination of the cartographers, changing shape at will and without warning. (Rushdie, 1989, p. 329)

The myths surrounding Southern California continue to exert a powerful grip on the way we see the region. In fact, they may actively be *preventing* us from recognizing many important changes that are occurring right in front of us! The challenge facing the urbanist trying to make sense of Los Angeles is to interpret the past, present, and emerging future landscapes of the region—to explain what this new "text" of the city means.

We find it useful to adopt some of the precepts of *postmodern urbanism* to explain what is happening in Southern California today.[2] The term *postmodern* is highly contested, and we have no wish to enter this debate here. Instead, let us simply note that we are using the term to convey a sense of *lost rationality,* that is, that the previous logics guiding the urban process have been altered, perhaps even irretrievably lost. In their place, we are discovering multiple rationalities that do not cohere into a single logic of urbanization. In this sense, postmodern urbanism is about complexity and difference. These are manifest in Los Angeles as an acute localization and fragmentation of social process. L.A.'s microgeography is increasingly fine grained and variegated, and the social heterogeneity and spatial extensiveness of the metropolis have encouraged intense and effective local autonomies. These appear in all walks of life, including politics, work, family, culture, and environment.

One important consequence of metropolitan sprawl is the difficulty of normal urban governance. Los Angeles may yet prove to be the harbinger of a new style of decentered politics (see Chapter 5 by Schockman and Chapter 6 by Hahn in this volume). The region is split into many separate fiefdoms, their leaders in constant battle. The problems of political representation include disputes between county and city governments, the resurgence of slow-growth and no-growth movements, and the difficulties associated with political participation by ethnic, racial, and cultural minorities. As the region continues to expand geographically, government has become increasingly remote and less able to respond to grassroots concerns. Consequently, formal and informal, legal and extralegal alliances have risen to press their claims (see Chapter 9 by Park, this volume). Gays, lesbians, gangs, feminists, and racial and ethnic groups work within the interstices of the formal power structures, which then become increasingly meaningless in the everyday lives of alliance members. Left to their own devices and encouraged by the rules of politics, elected officials exercise power within their fiefdoms in an increasingly autocratic and unaccountable manner. And so the bifurcation of formal and informal politics is intensified, the city seeming to operate on the edge of anarchy.

Figure 1.3. The postmodern world city? Los Angeles urbanism takes over the world.

In an apparent paradox, the rising preeminence of the local in the postmodern city has been facilitated by a global capitalism. The emergence of post-Fordist industrial organizations has resulted in an accelerated flow of global capital and an endless search for cheap labor supplies on an international scale, each having profound effects on the local economy (see Chapter 3 by Gabriel). At home, the consequences have been a rapid deindustrialization, especially in the Snowbelt, and (re)industrialization in the Sunbelt cities. Los Angeles, in perhaps a typically postmodern way, is experiencing both simultaneously. For example, the city recently lost the last vestiges of a major automobile manufacturing industry but also gave birth to the glittering towers of corporate hightechdom. Many communities have been irrevocably altered as a consequence of such changes (see Chapter 7 by Brill). Los Angeles is an "information city" with, at the same time, a proliferation of minimum-wage, part-time service industry jobs (e.g., fast-food outlets) and a massive informal economy (street vendors on freeway off-ramps, can-recycling efforts from the backs of trucks, etc.) powered by a seemingly unstoppable domestic and international immigration. The globalization of capitalism has connected the local ever more effectively to the worldwide developments of post-Fordism; what happens in downtown Los Angeles today may be the result of yesterday's fluctuations in international labor markets.[3]

In social terms, Los Angeles is now characterized as a "First World" city flourishing atop a "Third World" city. The latter term refers to the burgeoning population either engaged in the informal economy or paid poverty-level wages. They tend to be only marginally housed by conventional standards, or even homeless. The postmodern metropolis is increasingly polarized along class, income, racial, and ethnic lines (see Chapter 4 by Dymski and Veitch). The disadvantaged classes are overwhelmingly people of color. Their family lives are disrupted by the demands of a flexible, disorganized workplace (e.g., the pressure on both parents to work and the need for families to crowd together to be able to afford housing). These trends have been aggravated by a strong ideology of privatism, as well as the practical effects of privatization at all levels of government during the last two decades.

This acute openness to world trends would probably have been cushioned if it were not for an erosion of the linkages, horizontal and vertical, between branches of government. At home, the rhetoric of "less government" reflects a collective or civic aversion to dealing with social, economic, and political problems. So has the rise of "fiscal federalism" (federal transfers to California cities have declined by two thirds from their 1980 levels). Issues of community and the public interest have consequently taken a back seat. In effect, governments and populace have colluded in a decline of the commonwealth. The collapse of community is one reason why the postmodern city is increasingly without a credible infrastructure. Crime is rife. The drug culture is recognized as a rational response to the absence of mainstream employment. Health care for the poor is increasingly difficult to obtain (Chapter 12 by Tranquada). The public schools are in a shambles (Chapter 11 by Hentschke). Homelessness is pandemic in the region, and the welfare system is on the verge of collapse (see, e.g., Wolch, 1990; Wolch & Dear, 1993).

The apocalyptic images of the movie *Blade Runner* often seem anything but fictitious in Los Angeles. Air quality in the region is the worst in the country, despite increasingly draconian regulations. The physical expansion of the urbanized area has generated other acute, human-induced environmental crises, especially those connected with urban services such as water supply, toxic waste disposal, and sewage. These problems, together with the region's especially hazardous natural environment (earthquakes, floods, landslides, and fires), are proving increasingly intractable as the region struggles to maintain its economic impetus (Chapter 10 by Pulido). Although lip service is paid to environmental issues, the survival of nature in all its forms is a low priority. The intense localization, plus the absence of a conventional public transportation network, makes decentralization and diversity possible and even necessary to everyday life (Chapter 13 by Giuliano, this volume). Angelenos daily reinvent their city. For instance, there is no need to go downtown to enjoy the principal entertainment and cultural events of the postmodern city. There are alternative major theater districts in Pasadena, Hollywood, Long Beach, Orange County, and elsewhere. Art, music, and other forms of cultural expression flourish in the formal museums and performance halls scattered throughout the region, as well as informally on the sidewalks of East Los Angeles, South Central, and elsewhere (Chapter 8 by Boyd, Chapter 2 with Flick). Indeed, downtown Los Angeles is not *the* downtown for the vast majority of the region's population. Attempts have frequently been made to create a regional hub at the focal intersection of four major freeways. A large part of today's "typical" downtown agglomeration of commercial and residential high-rise is, however, as Mike Davis (1990) points out, a relatively recent and "perverse monument to U.S. losses in the global trade war" (p. 138)—which permitted a massive inflow of international capital for speculative real estate investment, backed by high levels of public investment.

Mapping Southern California's Urbanism

Los Angeles threatens . . . because it breaks the rules. (Banham, 1973, p. 236)

Is Los Angeles a model of 21st-century urban development (Chapter 14 by Fishman)? We cannot answer that question definitively, but two things are clear: (a) The Chicago school model is no longer valid in describing contemporary metropolitan evolution, and (b) Southern California can no longer be regarded as an exception to the rule. Los Angeles has always been a relatively decentralized metropolis, especially since the rise of the automobile. Through time, it has become even more powerfully polycentric, a characteristic that is intensified by the region's increasing social, political, and economic polarization. Indeed, the First World-Third World dichotomy is perhaps the principal signature of L.A.'s emergent urbanism. This dichotomy has been driven by the extraordinary openness of the region's economy to the forces of global restructuring and by the rise of the informal/underclass economy. It has been facilitated by the decline of formal politics and the collapse of the welfare state. The associated proliferation of informal local politics has exaggerated the culture of privatism that has been part of the political agenda since the 1980s.

Southern California is a place in which traditional modes of control are evaporating, and no single new rationality has yet appeared as a substitute. In the meantime, emergent forms of economic, sociocultural, and political relationships rush to fill the vacuum. The localization of these global effects is creating the new geographies of the postmodern city. In the chapters that follow, our contributors begin to outline a broad cartography of Southern California's emergent urbanism. We readily concede that they will not complete this task. But in their deliberations, they debunk many deeply rooted myths about Southern California; in the process, they also invoke some of the deepest, most difficult, and most urgent questions facing urban America. We hope that the voices raised in this book will begin to provide some answers about what is happening in one of our nation's most important cities.

Notes

1. The term *Los Angeles* is often used loosely in this book to refer to city, county, and region interchangeably. When specificity is necessary, individual authors will adopt the appropriate qualifier (*city, county,* etc.).

2. For a fuller discussion of the idea of postmodernism in the urban context, see the seminal work by Jameson (1991), as well as the articles by Dear (1986, 1988, 1989, 1991).

3. On the processes of contemporary industrialization in Southern California, see two volumes by Scott (1988, 1993).

References

Banham, R. (1973). *Los Angeles: Architecture of the four ecologies.* Harmondsworth, UK: Penguin.

Bartlett, D. (1907). *The better city.* Los Angeles: Neuner.

Calvino, I. (1972). *Invisible cities.* New York: Harvest/HBJ.

Comstock, S. (1928, May). The great American mirror: Reflections from Los Angeles. *Harper's Magazine, 156,* 715-723.

Davis, M. (1990). *City of quartz: Excavating the future in Los Angeles.* New York: Verso.

Dear, M. (1986). Postmodernism and planning. *Society & Space: Environment & Planning D, 4,* 367-384.

Dear, M. (1988). The postmodern challenge: Reconstructing human geography. *Transactions, Institute of British Geographers, 13,* 262-274.

Dear, M. (1989). Privatization and the rhetoric of planning practice. *Society & Space: Environment & Planning D, 7,* 449-462.

Dear, M. (1991). The premature demise of postmodern urbanism. *Cultural Anthropology, 6*(4), 538-552.

DeMille, C. B. (1925, August). Making the limits large enough. *Southern California Business,* 20-49.

Dykstra, C. (1926, July). Congestion de luxe: Do we want it? *National Municipal Review, 6,* 394-398.

Dykstra, C. (1941). Introduction. In G. W. Robbins & L. D. Tilton (Eds.), *Los Angeles: Preface to a master plan.* Los Angeles: Pacific Southwest Academy.

Fogelson. R. M. (1967). *The fragmented metropolis: Los Angeles, 1850-1930.* Berkeley: University of California Press.

Garreau, J. (1991). *Edge city.* New York: Doubleday.

Gilbert, R. (1964). *City of the angels.* London: Secker & Warburg.

Jameson, F. (1991). *Postmodernism, or the cultural logic of late capitalism.* Durham, NC: Duke University Press.

Kroeber, A. L. (1925). *Smithsonian Bureau of American Ethnology Bulletin 78: Handbook of the Indians of California.* Washington, DC: Government Printing Office.

Los Angeles County. (1922). *Proceedings of the First Regional Planning Conference of Los Angeles County.* Los Angeles: Regional Planning Commission.

McWilliams, C. (1949, October). Look what's happening to Southern California. *Harper's Magazine, 199,* 21-29.

Olmsted, F. L., Jr., Bartholomew, H., & Cheney, C. H. (1924). *A major traffic street plan for Los Angeles.* Los Angeles: Traffic Commission.

Rushdie, S. (1989). *Satanic verses.* New York: Viking Penguin.

Scott, A. J. (1988). *Metropolis: From division of labor to urban form.* Berkeley: University of California Press.

Scott, A. J. (1993). *Technopolis: High-technology industry and regional development in Southern California.* Berkeley: University of California Press.

Smythe, W. W. (1910, April). The significance of Southern California. *Out West, 32*(4), 288-302.

Soja, E. (1989). *Postmodern geographies.* New York: Verso.

Suburbia—exurbia—urbia. (1957, April 1). *Newsweek, 49,* 35-42.

Suisman, D. R. (1989). *Los Angeles boulevard.* Los Angeles: Los Angeles Forum for Architecture and Urban Design.

Viehe, F. W. (1981). Black gold suburbs: The influence of the extractive industry on the suburbanization of Los Angeles, 1890-1930. *Journal of Urban History, 8,* 3-26.

Wolch, J. (1990). *The shadow state: Government and voluntary sector in transition.* New York: Foundation Center.

Wolch, J., & Dear, M. (1993). *Malign neglect: Homelessness in an American city.* San Francisco: Jossey-Bass.

Wolfe, T. (1964, December 1). I drove around Los Angeles and it's crazy! The art world is upside down. *West,* 19-20, 22, 27.

Los Angeles in the 21st Century

STEVEN B. SAMPLE
President of the University of Southern California

I think we stand at a crucial moment in the history of Los Angeles. We have an incredible opportunity to build our city anew, not as it once was, but as it should be.

Representing Los Angeles

ROBBERT FLICK

IN CONVERSATION WITH

MICHAEL DEAR

Michael Dear: Robbert, the first time I saw your work, I immediately thought, "That's it, that's Los Angeles!" That's exactly how I see this city. From the street, and at speed, it's a punctual, linear experience, not continuous as in a movie. The city is perceived essentially as an interrupted sequence. In addition, your photographs convey the flatness of the city, the absence of verticality in Los Angeles, its unidimensionality. Are these some of the effects you're striving for in your work?

Robbert Flick: Yes, these are some of the core problems in representing Los Angeles visually. I've been wrestling with them since the late 1970s—how to photograph Los Angeles in a manner that seemed appropriate to the locale? What has always struck me about the city is the possibility of the scan, a left-right/right-left movement. There is always an open sky. The horizon line here is perpetually present in any vision of Los Angeles, and I think that has a lot to do with the shaping of the perceptions, the notions of light and space. This kind of openness, combined with speed, was the starting point for me. During the 1980s, I started more and more to see the horizon line being broken up. The scan was interrupted by the increasing number of tall buildings beginning to poke through. So I started to search for funding to photograph Los Angeles, mainly because I was concerned about the fracturing of the urban spaces.

As that effort progressed, I began to realize that my representation of Los Angeles was totally based on being in the driver's seat. So then I started to develop a series of visual devices that simulated the experience of driving, particularly the way you look down the roadway to anticipate what's coming next. You fix your gaze maybe half a block ahead and then as you drive closer, you skip to the next half-block. It really tickled me that you described it as an "interrupted" vision because that's exactly the perception I have when I'm driving. It's a hop and a skip, a hop and skip.

I tried to capture this perception from the car but it was impossible. So I then did it on foot. I walked Wilshire Boulevard, and I walked La Cienega. But it still wasn't feasible because of time involved—the distances were too great. There is this discrepancy between the perception of time and the actual passage of time. Then along came the computer and the video camera and all of a sudden, everything coalesced.

MD: I want to hear more about your use of computers, but let's just stick with some more questions of principle for the moment. You use the words *time* and *space* a great deal in descriptions of your work. Geographers, architects, filmmakers, and other artists are typically obsessed by space and its relationship with time. Can you be more specific about the ways in which time and space figure in your compositions?

RF: You've got to start from the notion that a photograph is an illusion.

MD: On another occasion, you called it a fiction.

RF: Right. It is a fiction; it is a construction. It's an illusion of a reality; then the illusion becomes a reality. The way time functions in a photograph is different from the way time functions in film, for instance. In a photograph, time is basically suspended or frozen, so the type of scrutiny that can take place is rather substantial, yet it is still dependent on what is in the photograph that generates or invites scrutiny. Of course, photography has a long tradition of orchestrating the visual in such a way that the viewer becomes involved. There are formal structures that you could use to set things up. There's also a long tradition governing how photographs function and perceptions are manipulated.

The trouble with all this is that photographic set-ups take time, and as we move toward the end of the 20th century, our perception of photographs is no longer ambulatory. Instead, it's essentially electronic and governed by the way in which the media have affected our way of seeing, so that the individual image is no longer recognized for what it contains but rather for what it refers to in the experiential apparatus of the viewer. Now, by setting up relationships between frames of still images, I begin to create tensions among the frames. Hence, before you encounter the iconographic content of a photograph (remember, the single photograph points you toward the icon, rather than its content), it's been subverted! When time is frozen in a photograph, a whole host of things happen. The perceptual vise that's created in my photographs has to do with what's in one image and what's in the adjacent images; this forces the viewer to begin the act of comparing, rather than the act of naming.

The experience of Los Angeles, whether on foot or in an automobile, is basically of texture. There's a certain sense of depth as you travel down the streets and start looking at things. The primary textures are basically the facades of buildings because so much of Los Angeles is flat. It's a lot like a studio back lot! The facades are either one or two stories high, and the adornment of the facades becomes an indicator of what might be inside. At that point, things really get interesting because what might be inside is perhaps on top of what was previously there. There's a kind of layering.

MD: As in archaeology?

RF: Yes, an archaeology, or perhaps a road cut. I'm very interested in the way those textures allow me and the viewer to begin a process of interpretation that is active rather than passive. It points to something and allows you to construct things.

MD: Would it be fair to interpret your reference to *texture* as a spatial construct?

RF: Textures are spaces. But I prefer to call them *ecologies*. For instance, in 1986, I did a visual inventory of Wilshire Boulevard. I identified roughly 32 ecologies, taking into account the type of visual experience of the spaces that were present. These pertained to the width of the street, the way in which the shadows fell, the spaces between places, as well as the literal textures.

MD: Is there anything particular about Los Angeles that has led you to these artistic concerns?

RF: Ah! I've always wanted to be here. I heard about Los Angeles a long time ago. I prefer it because in some curious way, it has no history.

MD: What do you mean?

RF: Well, I grew up in Amersfoort in Holland. The school I went to was started in 1450. The city has an incredibly pervasive sense of history. The same is true in parts of the East Coast of the United States. But when I went to Canada, what I liked about Vancouver was that history was almost not there—we hadn't messed with it yet. I could look around Stanley Park and still sense why the totems looked the way they do, just by looking at the cedars and the way in which the dead cedars lay on the ground. And I liked that. What I like about Los Angeles is that nonhistory.

MD: So you wouldn't be doing the same work if you were in, say, Chicago?

RF: No. For one thing, in Chicago there is a much clearer sense of community. Now obviously, I live in a community in Los Angeles, too—a community of artists. But in Chicago, there is a strong sense of this or that ethnic community. Or people say, I'm from the Southside. You lay claim to your community with great vigor. In Los Angeles, you may say you're from the Westside or wherever, but basically Los Angeles allows you to be anonymous. In Chicago, you couldn't—at least I didn't have that sense when I lived in the Midwest.

MD: Apart from questions of nonhistory and of no community, there's also the issue of landscape, isn't there? You mentioned, for instance, the flatness of Los Angeles. For me, one of the most powerful features of your work is the stark horizontality of the cityscapes that you've assembled. What's interesting is how the viewer's journey through the horizontal stream of a boulevard is often brought to a screeching halt by the encounter with a monumental image such as, for example, the Norton Simon Museum in your Rose Bowl Parade Route sequence. Here, the perceptual experience is spread over a dozen or more frames. But the viewer is also arrested by the small scale, as when one's attention is caught by the purely ephemeral, such as a street corner argument that emerges and dissolves through a sequence of only two or three frames. Or when a woman walks out of a single frame, and a passing truck interrupts your view.

RF: Actually, there is a connection here with the Midwest. It's ironic that you should stumble across that. I worked for 6 years in Champaign-Urbana in Illinois and basically fell in love with the landscape. It was there that I really started to address landscape, especially its flatness. For me, there is a direct correlation between the edge of a cornfield and the edge of a street. It may be just a private thing, but there is a connection. When I first came to Los Angeles, what I liked was that the landscape had no boundaries—just this layering, everything occurring simultaneously. There was always this possibility of things on top of each other—there's that archaeological metaphor again. It might have had something to do with what was physically present in the city at that time, during the late 1960s.

MD: Is there anything else that makes Los Angeles a special place for you?

RF: Oh, the anonymity. I have to come back to that. Total anonymity! But it's not simply that you can lose yourself in the city. It's more than that; things looks the same, but they never are. You go down the same street, and it's always different. Those huge suburban tract developments of the interwar years all have a similar feel to them, but they're really all different. And I'm very interested in that. I'm interested in the stuff that makes up a place—the stuff that you don't taste, that gets in corners. I like it because that's basically how perceptions are shaped.

MD: Let's talk more about how perceptions get shaped. How do you make your selection of images? I know that once you've made a videotape of the cityscape, you are deeply involved in selecting individual images and in arranging deliberate overlaps and juxtapositions to achieve an effect. Can you tell us how this composition process works?

RF: First, there's a selection process from the videotape. I analyze each frame. I look at pictures next to each other and pictures in conjunction so that they create context for each other. And I envision these in relation to a larger field, much as one might address the issue of creating space on a canvas. You could say that my selection of individual frames is based on visual content plus what affective qualities a particular configuration of frames will generate. Needless to say, there's also a sensibility about the history of photography as well as an empathetic sensitivity to what is unfolding in front of me. The whole vision comes together simultaneously, bringing with it a wonderful sense of reality. There's also an element of voyeurism involved. I scrutinize; I stare at a photograph. It's a kind of "scopophilia" on my part, I suppose.

MD: Do you deliberately try to trick the viewer? Or surprise the viewer in any way?

RF: Oh, there has to be some involved! I'm always anticipating the spectator's response. You begin with the logic of the street. Then you consider how to display it visually; the frozen matter of the photograph is different from the linearity of the videotape. The landscape is all there, but I create a fiction out of it because I can repeat things. I can shorten spaces or emphasize certain textures. It's like making music or telling a story, even though I don't know precisely how it's going to turn out.

MD: Let's follow that analogy through for a moment. On one hand, you say that you select and compile a sequence of individual images that are obviously related in some way, like words and chapters in a book, or notes and lines of music in a composition. But on the other hand, what about the *overall* narrative? All these words and notes, all your photographs, must ultimately add up to a coherent narrative if they are to succeed. So how do you make the leap from the individual image, or set of images, to the narrative as a whole?

RF: I used the word *ecology* earlier. Basically, I strive to contain the overall sense perception of an area within a single field. I try to encompass both entries and exits to this ecology and to foreground what is most manifest in this set of encounters. For example, along Melrose, I will orchestrate one set of visuals. But if I then go a couple streets north or south, the textures are different, and I have something new to play with. The overall shape of the narrative is really dictated by how big the piece is going to be, as well as what I want to make manifest from the experience. But obviously, I don't come into a place as an innocent! I've selected the area, studied it, and set up a shooting strategy to make manifest what I believe is dominant.

MD: But, still, how do you come to closure about a place? Clearly, in one sense, a place narrative can never be finished, but how do you decide when a story is ended? Crudely stated, is it simply a matter of how many images you can squeeze onto a single page?

RF: Crudely stated, page size is one important artistic constraint. But generally speaking, much, much more is involved. Different places have different textures; they demand different strategies of containment.

MD: Let's turn now to some of the more technical aspects of your work. When I first came to your studio and watched you work, I was immediately impressed by the high-tech aura surrounding what you do. I expected to find you "starving in your artist's loft," surrounded by the clutter of your trade. Instead, you sat in a neat, almost sterile computer lab. I had the impression of a cold, almost clinical approach to creativity. How do you see yourself in this technical revolution? Is yours a virtual reality? Are you lost in cyberspace?

RF: It's just a—I wanted to say computers are just a tool, but they're not. There's something else at stake there. I think it was Sarkowski who said that when photography was invented, everybody knew what it could do, but nobody knew what it was for. I have a similar sentiment about the computer. Yes, at one level, it's a scientific miracle. But I am not a scientist; I don't even know how to program the machine. On the other hand, I am pushing what is available to its limits.

MD: You said earlier that when the computer came along, everything fell into place. Your current work wouldn't be possible without such technical innovation?

RF: Absolutely not. There are vast numbers of images to manipulate and infinite numbers of permutations in the overlaps between them. I couldn't do it without the machine.

The pieces that I did during the 1980s were 100-, 200-, and 300-frame compilations of individual photographs that were physically painful to complete. I would spend 15 or 16 hours straight, standing on a concrete floor in a darkroom. It was physically very demanding, but there was also the problem that my mind was going so much faster than my hands. I'd make my selection after the shooting and begin to put things together. Can you imagine thousands of little pictures, all of them collated in a rudimentary way? Then I faced the task of articulating them in a way that people could experience. That was just horrific!

I don't mean to say that working with a computer is effortless; there's still an incredible amount of time and effort involved, but it's a different kind of effort. My attention is no longer distracted by the physical pain of making it something, so now I can stay focused on the artistic concept.

MD: You mentioned the history of photography earlier, and that made me think about influences. When I first encountered your apparently repetitious streetscapes, I was reminded of Andy Warhol or some of David Hockney's collages of Polaroid snapshots. Are there echoes of Warhol and Hockney in your work? What artistic influences brought you to where you are today?

RF: Much of what I do is rooted in conceptual art. Of course, there's Ed Ruscha with his Sunset Strip and gasoline stations—very important work. Then there's Robert Venturi and Scott Brown's *Learning From Las Vegas*. But there is so much that one is familiar with! The difference is that I approach my work from a photographic base. The questions I am asking have to do with the nature of photographic representation, whereas Warhol and the others simply used photographic representation in their work. There's a subtle difference here in the way that Hockney puts his images together and the way that I assemble mine. Hockney's are rooted in a pictorial space that is more related to the act of painting and to the beginning of cubism. Mine is much more a *narrative* in the sense that I'm deliberately interrogating juxtapositions.

MD: You once told me that you are involved in the creation of conditions for memory to occur. That struck me as a provocative, even poetic phrase. Could you explain it?

RF: What I meant is that when people encounter an image of a familiar place, they will say, "Oh, this is so and so." So when I present a field of images that's labeled (say) "Pasadena" or "Rose Parade," people who know that place will automatically gravitate toward it, seeking out the familiar. Those who don't know it will instead seek out some dominant characteristic that they can identify. It's in the process of focusing—that's where memory can occur. The moment that recognition unfolds, memories are being jogged, and whatever that person is looking at is viewed through the filter of his or her own experiences. It's exactly at this moment that the picture becomes complete, if only for an instant. That is how I set things up, and that's why I call it fiction.

MD: Let's come back to the fictions of Los Angeles. Tell me more about the stories you are trying to communicate to us.

RF: I have been looking hard at Los Angeles for almost 20 years now, perhaps even longer. I've lived here since 1976, but I first came to the city in 1967. My love for the place has always been with me. I started with a simple idea that there were certain locations in Los Angeles that were identifiable in economic, ethnographic, and demographic characteristics—for example, Pico-Union, Broadway in downtown L.A., the financial district, Echo Park, East L.A., and Boyle Heights. And of course I knew about the Indian communities on Artesia Boulevard, the Vietnamese communities in New Westminster, and so on. But I didn't really know them. I'd go there to shop perhaps, and I'd look at them hard, but not willfully. Once I started to look at them with purpose, however, they just broke apart.

MD: What do you mean?

RF: In the sense that I thought I could handle Los Angeles. But no way! Night and day, you're stuck in this maelstrom of people. It's immovable; you can do nothing with it. Sometimes I feel so detached from it, almost as though I am nonexistent. It's in this sense, too, that Los Angeles broke apart for me.

What interests me is how all this diversity and density can coexist. I used to follow the east-west patterns, such as Imperial and Valley Boulevards, to get a sense of the evolution of the city. But now I prefer going north-south, along Alvarado, for example. When you start at Echo Park and go down Alvarado, within a distance of less than a mile and a half, you experience a reshaping. The history of Los Angeles plays itself out before you, just in this short space: from empty places to funky houses with lots of space between them, then that compaction and incredible density along Sixth Street, before the strange, manicured greenery along Pico-Union.

MD: I hesitate to ask a photographer this question, but what *words* would you use to describe your narratives of Los Angeles?

RF: I would say *topographic,* but it's more than that—*layered,* perhaps. *Trajectory* is an evocative word. It's reminiscent of when I was a kid. I wanted to be a biologist. We would place a meter-square frame on the ground and discover every scrap of life within it, and that one-meter sample would come to stand for the whole. You had to inventory the fungi, the grasses, the flowers, the pH content of the soil—all of that. There is some of this in my photography, too—a taxonomic urge, if you like, a need to take stock.

MD: That's curious! The scientist in you seems to be searching for order and comprehensiveness. But the artist in you concedes that understanding is continually slipping through your fingers and eluding your attempt to grasp it.

RF: This is an amazingly accurate perception. When I started out as a photographer, I quickly discovered that I couldn't approach photography as if I was a hunter. Only when I located myself in a different space could I learn from it—a kind of detached yet empathetic mode is vital in the process of my work.

The Inner-City "Problem"

PAUL C. HUDSON
President and Chief Executive Officer, Broadway Federal Savings and Loan

Let me tell you about my drive over here. I started out at Midtown Shopping Center at Venice and La Brea, which was burned out in April 1992 and is still not rebuilt. We lost a 30,000-square-foot grocery store, a Boys Market, a check-cashing operation and cleaners, all not rebuilt in the past 2 years. I went down Crenshaw past Crenshaw and Adams, where two corners remain burned out from April 1992. I proceeded down Crenshaw to Martin Luther King; two corners at MLK and Normandie are burned out. I then went down Vermont to Florence. Vermont is just a disaster—every block has lots that have been burned out. At Florence, I took a left to see one of our REOs, a 15-unit apartment building at 140 East Manchester that we have been unable to sell. When I took a right turn to return to USC, I saw two cars parked in the middle of the street, and then two brothers came up on bicycles. I did not want to cause trouble; I wasn't going to panic. As I left, what struck me more than my panic was how many brothers in that block were out on the street at 3:30 in the afternoon, clearly without meaningful employment. This is a sampling of many neighborhoods in the South Central: men, 25 to 45, not working, hanging with their buddies. Things have not improved in the past 2 years; there has not been a significant rebuilding effort. There has been no leverage from project to project. Strip malls are in place, but the kind of job-generating projects that Rebuild Los Angeles imagined are not. This is not simply due to the recession. There is a different economic reality in inner-city Los Angeles.

Remaking the Los Angeles Economy 3
Cyclical Fluctuations and Structural Evolution

STUART A. GABRIEL

Recent years have witnessed a moderate rebound in economic growth in Los Angeles, in the wake of the severe downturn of the early 1990s. The upswing in activity, however, has occurred in the context of a costly and continuing restructuring of the L.A. economy. Although the county contained about one half the statewide manufacturing jobs in 1970, this share declined to 41% of the state total in 1989 and to about 36% in 1995. The unemployment rate in Los Angeles County rose from 5.8% in 1990 to 9.7% in 1993; during that same period, the number of unemployed people in the county moved up by about 165,000 to almost 450,000 persons.[1] In addition, the county's welfare caseload mushroomed during the early 1990s, and research shows that the majority of the growth resulted directly from the decline in manufacturing employment.

Although economic indicators for recent years suggest an expansion in economic activity—both for Los Angeles County and for the state of California as a whole—growth remains selective in both location and economic sector. For instance, by late 1995, little recovery was evident in major portions of the real estate industry; similarly, the sizable job losses evidenced in such sectors as aerospace and defense contracting appeared to be largely permanent. In contrast, some strengthening in economic activity was recorded in other sectors, including entertainment, telecommunications, apparel, light manufacturing, and international trade (Table 3.1). The upward movement in economic activity for the L.A. and California economies occurred even as the national and international economies

Table 3.1 Selected Employment Trends, Five-County Area, 1980-1989

Industrial Classification	Percentage Change in Employment, 1980-1989	Number of Workers (1989)
Mass Production Industries	**-46.3**	**17,162**
Petroleum and coal products	-5.2	8,811
Tires and inner tubes	-87.0	235
Blast furnace and basic steel	-72.5	3,688
Motor vehicles and car bodies	-40.4	4,428
Craft Speciality Industries	**20.3**	**334,748**
Apparel	15.1	108,024
Furniture and fixtures	1.9	51,149
Printing and publishing	30.5	88,659
Leather/Leather products	-53.8	5,069
Jewelry, silverware, and plated ware	19.7	3,069
Motion picture production and services	49.6	78,778
High-Technology Industries	**-6.1**	**362,215**
Office and computing machines	-37.9	23,205
Communication equipment	27.6*	17,041
Electronic components and accessories	-6.8	52,122
Aircraft and parts	7.0	127,121
Guided missiles, space vehicles, and parts	47.2	80,447
Instruments and related products	-3.1*	62,279

SOURCE: Adapted from Scott, 1993, Table 2, p. 185.
*1980-1987 figures.

slowed; in 1995, the rate of nonfarm employment growth of the California economy was about 2%. The rebound in statewide economic activity occurred in the wake of the longest period of state economic weakness since the 1930s.

Widespread media commentary and policy debate surrounded the early 1990s downward spiral in Los Angeles economic activity. In this chapter, I examine some of the causes associated with the recent economic recession in Los Angeles. That discussion is followed by a series of policy proposals that focus on the long-term economic viability and growth of the region. In addressing these issues, I invoke an important distinction between cyclical fluctuations and more permanent evolution in the structure and composition of the regional economic base.

Structural Evolution

The substantial weakness evidenced in the Southern California economy during the early 1990s derived from a combination of cyclical downturn and structural economic change. The structural evolution of the L.A. economy is a continuing process prompted by such factors as the relatively high costs of local production, regulatory restraints thereon, and changes in the magnitude and spatial incidence of federal government aerospace and defense spending. These factors have resulted in secular declines in heavy industrial manufacturing in the L.A. region; for example, automobile assembly, furniture, and plastics manufacturing have largely moved away. Also, realignment and cuts in federal defense and aerospace budgets have served to significantly reduce employment and output in those traditionally important sectors of the local economy. In large measure, those federal spending cuts result in structural unemployment, whereby a difficult mismatch is created between the background and skills of highly sophisticated former defense industry workers and the often lower-skill requirements of newly emerging jobs. Short-term policy fixes are difficult here, in that few of the laid-off aerospace and defense workers can expect to return to their jobs in Southern California. Worker retraining and worker moves (to other areas or employment sectors) may offer some relief to the structurally unemployed; both strategies, however, are difficult to orchestrate and involve substantial costs to the displaced workers. Accordingly, the unemployment problems of this group will likely persist for some time because defense-related output remains significantly below previous highs.

Cyclical Fluctuations

A variety of cyclical factors have been equally critical in the recent downturn of the Southern California economy. The list here includes substantial overbuilding in various real estate markets, aided in part by a regulatory environment that created serious moral hazard on the part of lenders and allowed their less than full attention to bottom-line real estate economic fundamentals. An oversupply of office space remains to this day in many areas of the city, due to overbuilding and reduced demand for space. Similarly, a "bursting of the house price bubble" led to sizable declines in housing investment returns and reduced demand for housing. Because of a lack of availability of residential construction funds and substantially reduced rates of new construction during much of the early 1990s, however, a firming of that market appears to be under way. The widespread and sizable losses of real estate equity spilled over into reduced levels of consumer confidence and consumption spending as well and served to dampen the demand for luxury automobiles and other consumer durables. Other factors contributing to the cyclical downturn in the region included substantial retrenchment and loss of employment in the financial services industry; for both banks and thrifts, the demise resulted in no small measure from losses incurred because of unprofitable or defaulting real estate loans. The generalized national and global nature of the cyclical downturn of the early 1990s, as evidenced by reduced levels of economic growth among major U.S. domestic and international trading partners, similarly resulted in diminished demand for Southern California goods and services.

Charting the Economic Future

Having identified certain cyclical and structural components of the Los Angeles economic downturn of the early 1990s, I now turn attention to the necessary preconditions

Table 3.2 Population of Los Angeles County (Percentage)

	1970	1980	1990	2000*
Anglo	70.9	52.4	40.8	31.0
Black	10.8	12.7	11.2	9.8
Latino	14.9	26.1	36.4	44.0
Asian	3.4	8.8	11.5	15.2
Total (%)	100	100	100	100
N	7,032,075	7,477,503	8,863,164	10,924,000

SOURCE: Adapted from Ong, 1993, Table 1, p. 33.
*Projected.

for long-term economic revitalization and growth in the region. First, because of its location at the edge of the Pacific Rim, Los Angeles is characterized by some *geographic comparative advantage* in East-West international trade. The Los Angeles Customs District recently emerged as the busiest in the nation, surpassing New York in dollar volume of trade. The geographic comparative advantage in East-West trade is enhanced by the vast diversity of population groups in the city (Table 3.2). For instance, the Asian and Latino populations of the city possess the language skills, cultural and business insights, and family ties necessary to the development of successful trading relations with Asian and Latin American business partners. In the wake of NAFTA, those trade relations are increasingly trilateral, as northern Mexico becomes an important point of assembly and production for imports to the United States from Asia. The L.A. region likely will contribute importantly to and derive significant benefit from the anticipated rapid growth in Pacific Rim and Latin American trade in future years. Local public policy should focus on the leveraging of local comparative advantage to further enhance the national leadership role and local economic multiplier effects associated with that trade.

As is evident to even the casual observer of the Southern California economy, local real estate prices have moved down significantly from their peak levels of the late 1980s. Those declines have been particularly pronounced in residential, office, and retail markets; property values in some market segments have moved down by more than one third from their late 1980s peak levels. Undoubtedly, the *sharp downturn in real estate markets* has been the cause of some pain and dislocation among homeowners and businesses in Southern California. The substantially reduced wealth levels of many homeowners—because of the reduction in home equity—resulted in damped levels of consumer spending that in turn further exacerbated the generalized economic downturn. The combination of overbuilding in the office sector and reduced demand for office space—due both to regional economic weakness and to technological innovation and new efficiencies in the use of office space—served to push down cash flows associated with office investments as well as the valuation of office properties.

The severe retrenchment in real estate valuations, however, makes for significant *improvements in the relative affordability of residential and nonresidential space* in

Table 3.3 Poverty in Los Angeles County (Percentage of People Below Poverty Line)

	1969	1979	1989	N (1989)
Anglo	7.8	7.5	6.6	229,000
Black	24.2	23.2	21.2	203,000
Latino	16.6	19.2	22.9	744,000
Asian	11.2	14.6	13.2	125,000
Total	753,000	985,000	1,301,000	1,301,000

SOURCE: Adapted from Ong, 1993, Table 4, p. 38.

Southern California, and in so doing, similarly provides a necessary ingredient to the long-term economic competitiveness of the region. In 1988, for instance, a mere 12% of households in Los Angeles could afford the median-priced home in the county, which sold for well over $200,000. In contrast, during that same year, about 50% of households nationwide could afford the median-priced home, which sold for about $100,000. By mid-1995, largely because of declines in house values, nearly 40% of households in Los Angeles could afford the median-priced home in the county. At the same time, rents on class A office space in the downtown L.A. market were highly competitive with those of other major domestic and international markets. Affordability of housing and office space is an important determinant of firm location choice and job growth; the increasingly competitive nature of property markets in Southern California will serve to retain households and jobs as well as to encourage movement into the area of new households and firms. Without such improvements, Southern California would have witnessed more attrition of jobs to areas of cheaper housing and production costs.

A further important ingredient to the future vitality of the L.A. economy is the provision of economic opportunity to the broad spectrum of population groups and communities. As indicated in Table 3.3, minority groups in Los Angeles County suffer persistently high rates of poverty. Data further indicate sizable racial and neighborhood disparities in homeownership and in mortgage loan origination. Damped rates of homeownership among minority households and neighborhoods result in lower levels of wealth accumulation, diminished housing turnover rates, and problems of neighborhood stability. Also, there not only disproportional low representation of financial institutions in inner-city neighborhoods, but also a low incident of supermarkets, entertainment facilities, retail outlets, and insurance services.

Numerous private market analyses and public policy studies indicate that the *revitalization of inner-city areas* would generate a host of private and social benefits. Long-standing underservice of neighborhoods by supermarket chains and other retail establishments suggests the potentiality of economic profits to entrepreneurs who become active in those markets. Redevelopment of inner-city neighborhoods would create jobs, raise property values, and generate sales and property tax revenues. Some of the increment in property tax revenues could be used to enhance local public services and infrastructure in a manner consistent with the goal of revitalization.

Another critical ingredient to the long-run economic viability of the region is significant *improvement in the quality of public education.* Public schools in Los Angeles County vary considerably in their ability to deliver high-quality educational services; schools within a short geographic proximity of one another in areas of downtown vary from the top to the bottom tier of schools in the state according to standardized student test scores. Investment in human capital and human resources remains a critical precondition to long-run economic growth. Such a notion is of particular importance in a metropolitan area as ethnically and racially diverse as Los Angeles. The availability of a high-quality and highly trained workforce is an important determinant of firm location choice; investment in the preparedness and competitiveness of the L.A. workforce undoubtedly augurs well for the long-run vitality of the region.

If high-quality public schools are of critical importance to household location choice, the reverse is also true—that significant deterioration in the quality of public education has precipitated out-migration of population from the city to outlying and more distant areas. *Investment in training and human resources* will significantly aid in job growth through the creation of small business enterprises and the advancement of entrepreneurial activities. Indeed, such indigenous and small-scale entrepreneurial activity is already a strength of the L.A. economy. Accordingly, financial support of schools, colleges, and universities—and critical assessment of the quality of educational services delivered—is of the utmost importance to the future viability of the region. The State of California Superintendent of Public Instruction recently offered to substantially reduce bureaucracy and enhance regulatory flexibility among select school districts that provide evidence of new and significant progress toward measured educational achievement. School districts in Southern California may indeed want to participate in that program and, in so doing, work to develop innovative approaches to the delivery of high-quality public education.

The Southern California region has succeeded through the years in attracting a large number of highly trained professionals to work in the aerospace and defense industries. Not only have those industries lost substantial numbers of jobs in recent years, but some further consolidation of employment is expected. Local public policy should strive to *leverage this highly qualified workforce into new business sectors and activities* and thereby retain those households in the area. Such evolutionary activities are undoubtedly difficult to start up and occur only incrementally. Yet having said that, new public-private partnerships, such as those for the development and manufacturing of electric or hybrid automobiles, offer some growth potential. Numerous university and public policy research units are currently evaluating strategic new directions for the Southern California economy; those efforts seek to match existing local economic resources, skills, and areas with promising sectors of economic growth.

After assessment and identification of high-potential areas of L.A.'s investment and growth, a local *industrial policy* would be useful. It would include new state and local government incentives to enhance private investment, production, and marketing in economic sectors of local comparative advantage. Specifically, an industrial policy would take the form of an overarching strategic plan to promote output and employment in specific economic sectors and areas and would include such policy tools as tax abatement, local enterprise zones, and other incentives for firm location and hiring. Such a policy could also include provision of low-interest business loans and development of venture capital funds to aid in start-up of production, establishment of a consortium to facilitate the marketing and sales of specified products, and the like. Industrial policy is central to a strategic and interventionist approach to the transformation of the L.A. economy, whereby government would work hand in hand with the private sector to effectively "prime the pump" in identified high-potential sectors of economic activity.

The attractiveness of the California dream has been further diminished by a combination of economic and *quality-of-life* factors. Our work at the Federal Reserve Bank of San Francisco has demonstrated the sensitivity of Californian domestic migration flows to both economic and quality-of-life factors. Although economic factors explain a large majority of the net out-migration of domestic population from California during the early 1990s, some importance is also attached to the diminished attractiveness of local amenities. In particular, perceived problems of crime, gridlock, air quality, and so on have dampened the migration to the area of households and firms; similarly, these problems have led to the out-migration of L.A. residents, both to the far reaches of the metropolitan area and beyond.

Household and firm location choice are especially sensitive to issues of public safety. All evidence suggests that jobs and households will often move from or otherwise avoid areas plagued by problems of crime, gangs, and other threats to public safety. Continued deterioration of local public safety in many areas of Los Angeles may serve to seriously threaten the future economic vitality of the city. Public and private *investments to achieve high levels of public safety* are therefore critical to the long-run economic viability of Los Angeles. Inner-city redevelopment, like economic revitalization of the region as a whole, requires a host of initiatives—both public and private—to ensure the safety of ordinary citizens on the streets of the city.

Mild weather and beautiful beaches were among the amenities that brought large numbers of households and jobs to the L.A. basin. By the same token, renewed attention to issues of clean air, clean water, and attractive parks and beaches is also critical to the attraction and retention of jobs and households in the area. Although a strict air quality control policy may cost some jobs in the short run, there is little doubt that an aggressive antipollution policy—by helping to enhance the attractiveness of the region as a place to live, work, and vacation—is an important precondition to the long-run economic vitality of Southern California.

The city and county of Los Angeles must learn from other great metropolitan areas regarding the development of *alternative modes of transportation* to compete with the private automobile. As currently planned, the transportation system will consist of a variety of interlocking transportation modes, including bus, light rail, subway, and commuter rail. Regardless of final configuration, the objectives of the system—to elevate household and worker mobility and reduce commute costs, improve economic productivity, and otherwise enhance the environment and the quality of life—all figure importantly in the plan for a more economically viable and more attractive Southern California.

Conclusion: The Way Ahead

A multifaceted, strategic approach is required to ensure the necessary conditions for the long-term economic growth and vitality of the L.A. region. Such an approach, by definition, involves focus on the primary economic and quality-of-life determinants of household and firm location choice. Los Angeles is characterized by a few immutable attributes, notably including its geographic comparative advantage as a major gateway to trade with the Pacific Rim and Latin America. In addition, the city possesses the diversity of population that is instrumental to the consolidation and expansion of trade relations. To some degree, however, the locational advantages and amenities of Los Angeles have eroded through time by the relatively high cost and regulation of production; an ailing and inadequate transportation infrastructure; degradation of air, water, and other environmental resources; low levels of public school funding and educational attainment; and problems of public safety.

Those concerns require the intense and continuing focus of public and private initiatives, in the form of differentiated approaches to and new investment in educational services and local human capital, innovations and improvements in public transportation infrastructure, streamlining of unnecessary and cumbersome regulation, strict adherence to environmental quality goals, and public-private partnerships to rid neighborhoods of organized crime and violence. Strategic economic analysis and implementation of a local industrial policy would further serve to motivate private enterprise to invest and produce in identified areas of high economic potential.

Undoubtedly, there are trade-offs among newly proposed and existing policy initiatives. The streamlining of unnecessary regulation, for instance, should not be at the expense of environmental policy and educational achievement goals. Similarly, new initiatives—such as those to make neighborhoods safe from violence and to improve the quality of local public education—require significant expenditures. In some cases, those costs could be defrayed in part by private citizen and corporate involvement, user fees, and the like. Ultimately, however, difficult choices must be made regarding both the reallocation of scarce existing public funds and the development of new public funding sources. Undoubtedly, local government initiatives have been severely hampered in the wake of California's 1978 property tax limitation, known as Proposition 13. Although local tax increases are anathema to many, those tax costs should be carefully weighed against the potential areawide benefits of an economically revitalized and higher quality-of-life metropolitan area.

Finally, efforts must be made to ensure an inclusive economic system that provides a level playing field for all Southern Californians. Such an environment is critical to the efficient operation of the decentralized economic system as well as to the attainment of high levels of economic and social welfare. It is only in the context of the fair and open functioning of the economic system—in a manner that confers vested interests on all groups—that we can hope to secure the long-term economic strength and viability of the region.

Note

1. The employment data provide little insight as to the number of discouraged job seekers who dropped out of the labor force or to the number of workers who were involuntarily underemployed.

References

Blomquist, G. C., Blager, M., & Hoehn, J. (1988). New estimates of the quality-of-life in urban areas. *American Economic Review, 78,* 89-107.

Bowman, D., Ellwood, J., Newhauser, F., & Smokusky, E. (1994). Structural deficit and the long-term fiscal condition of the state. In J. J. Kirlin & J. I. Chapman (Eds.), *California policy choices* (Vol. 9, pp. 25-50). Los Angeles: University of Southern California.

Gabriel, S. A. (1995, October). California dreamin': A rebound in net migration? *Weekly Letter, 95*(33). (Federal Reserve Bank of San Francisco)

Gabriel, S. A., Mattey, J. P., & Wascher, W. L. (1995). The demise of California reconsidered: Interstate migration over the economic cycle. *Economic Review, 2,* 30-45. (Federal Reserve Bank of San Francisco)

Mattey, J. P. (1994, December). Effects of California migration. *Weekly Letter, 94*(43). (Federal Reserve Bank of San Francisco)

Ong, P. (1993, March). Diversity or divisiveness? In A. J. Scott (Ed.), *Policy options for Southern California* (Working Paper #4). Los Angeles: UCLA Lewis Center for Regional Policy Studies.

Scott, A. J. (1993, March). The new Southern California economy: Pathways to industrial resurgence. In A. J. Scott (Ed.), *Policy options for Southern California* (Working Paper #4). Los Angeles: UCLA Lewis Center for Regional Policy Studies.

Economic Referendum

JACQUELYN DUPONT-WALKER
Ward Economic Development Corporation

Efforts to live, relate, and work cross-culturally have intensified for those people who have felt left out. We must be sure that those of us who maybe didn't feel left out before join in that effort. The bicultural dialogue has to become a multicultural action group. The Black-Korean dialogue is now the Black-Korean-Latino dialogue. Other groups are beginning to buy property and to provide jobs. The economic referendum of April 29th, 1992, has shown signs of having some laudable successes. In fact, a magazine has even come out that's called *Good News*. It's needed because many of our publications say it is not journalistically responsible to tell good news. It is only journalistically responsible to dissect and analyze anything that appears to be good to prove that it really couldn't be that good. But *Good News* is here to forecast as well as to report that which is good to encourage us and empower us.

The question I have asked myself recently is this: Is the riot recovery taking too long? I don't believe that it is taking too long. I don't believe that we have arrived, despite that we've done better than New York, Detroit, Miami, and St. Louis. I believe the challenge to us if we're to move Los Angeles into the 21st century is to embrace and to respect the diversity that we have there—not treat it as an analytical fishbowl, but to get in and treat it as our own backyard.

Financing the Future in 4 Los Angeles
From Depression to 21st Century

GARY A. DYMSKI

JOHN M. VEITCH

> Los Angeles is an extreme example of a metropolis with little public life, depending mainly instead on contacts of a more private social nature. . . . It is embarked on a strange experiment: trying to run . . . a whole metropolis, by dint of togetherness or nothing.
>
> *Jane Jacobs, 1961, p. 73*

Los Angeles: Myth, Exception, or Paradigm?

For a century, Los Angeles has been for many the manifestation of the American dream. Although other cities grew incrementally through decades, Los Angeles emerged through riotous bouts of speculative excess. Other cities developed in concentric rings around central cores; Los Angeles splashed crazy-quilt across basins and canyons. Southern California's prosperity rebuked more sober regions, enticing a steady stream of migrants, and their wealth, westward. Two intertwined myths enveloped the city: Los Angeles was a land of preordained and uniform prosperity, and it was an Arcadia for white settlers, where other races existed only as street names and subordinate laborers.

AUTHORS' NOTE: We gratefully acknowledge the financial support of the Rosenberg Foundation, the Economic Policy Institute, the University of San Francisco, and the University of California, Riverside. This chapter could not have been written without the superlative research assistance of Brent de Ruyter, Michael Figueroa, Joon Park, Claudia Perez, Mirna Saab, Tony Sison, Travis Watson, and Betsy Zahrt, and especially the efforts of Mwangi Githinji.

Los Angeles once appeared to be an historical exception to the urban rule; the steady growth of similar urban structures throughout the United States and the world, however, has forced a rethinking of the rule. Edward Soja's (1989) celebrated chapter "It All Comes Together in Los Angeles" in his *Postmodern Geographies* argues that Los Angeles defines the new urban paradigm because it is a fulcrum for three global trends: accelerated immigration, dispersed production, and the internationalization of the division of labor. Further, Los Angeles has a special global role: "A growing flow of finance, banking, and both corporate and public management, control, and decision-making functions have made Los Angeles the financial hub of the Western USA and (with Tokyo) the 'capital of capital' in the Pacific Rim" (p. 192). Ironically, these trends turn the older mythology of Los Angeles inside out—it has become a city with a majority of minority residents, with extremes of both wealth and poverty.

This chapter examines the financial aspects of L.A.'s growth, from the Depression to the present. The evolution of banking relations provides a lens for understanding L.A.'s shift from economic exception to urban paradigm. The city's financial dynamic—its unfolding geography of wealth and race—offers a valuable vantage point on urban growth of the financial structures that have generated the expenditure flows that literally put the city on the map. Past financing arrangements define how the city's built landscape will be remade in the future. Too little or inadequately coordinated financing yields an incompletely realized geography. The observed city is the legacy of a structure of capital.

Through our financial lens, we find that Los Angeles differed for so long from other U.S. cities because of the boom-bust character of its development. Capital inflows have always generated the preconditions for L.A.'s economic growth; spatial and job expansion has simply followed. Increasing capital mobility and the internationalization of these capital flows are now beginning to affect other cities in the same way that capital flows have affected Los Angeles since its earliest beginnings.

Los Angeles thus foreshadows the impacts that international capital flows will increasingly have on urban development in the future. We show that for Los Angeles, globalization has led to deregulated financial structures that widen wealth differentials and spatial inequality. Los Angeles has not become Soja's global financial center directing capital flows throughout the Pacific Rim; it remains a locus of financial inflows, now from Pacific Rim immigrants and offshore banks. These capital inflows have sustained L.A.'s growth, at the price of deepening its social and racial polarization. The question is whether the city's economic and social environment is resilient enough to support future expansion.

Booms, Busts, and Big Capital

What has made Los Angeles unique as an economic metropolis is its virtually unchecked importation of capital and labor from elsewhere in the United States and the world. The opening of the transcontinental railroad in 1876 (in Los Angeles) made Southern California accessible to the East and Midwest (McWilliams, 1973, p. 114), just as dispossession and genocide had eliminated the region's previous (Native American and Mexican) owners. The way was clear for the prolonged promotion of the Golden State.

The growth dynamic underlying the development of Los Angeles has been the contradictory economics of the boom. Immigrants, initially from the Midwest, were lured by cheap housing and the promise of Arcadia, both constructed by real estate speculators. These newcomers were normally prosperous farmers or merchants who arrived with their wealth, validating speculators' financial advances. In addition, the immigrants brought their labor supply. Labor supply continually outran labor demand, keeping wages low. This allowed speculators to keep the price of additional new housing low and ensured an unabated flow of migrants. Although speculators prospered, the growth dynamic implied that residents could never establish stable employment relations at secure wages. Residents remained poor, the incentives for immigration remained in place, and, parcel by parcel, the agrarian paradise became a sprawling network of interlinked residential developments—a protocity.

A boom region continually imports financial capital. It is thus able to live beyond its means—to undertake more investment and consumption than its income from production can support. Capital inflows and increasing economic activity put upward price pressure on desirable areas inside the urban fringe. Thus, as the boom proceeds, real estate price "bubbles"—phases of rapid asset-price appreciation—emerge and spread. Once launched, bubbles often persist because households and firms can use paper gains on existing assets to acquire yet new assets. A boom economy may eventually become a bubble economy, in which "sound" and "speculative" development become indistinguishable.

Bubbles continue to feed the boom but make its economic foundations increasingly tenuous. The bubbles raise asset prices in desirable areas. Consequently, an ever smaller portion of the region's population can afford homes in these areas. Capital inflows become more and more crucial in sustaining the boom. Meanwhile, the lower-income workers building the metropolis are forced into less desirable inner-city areas, increasing asset-price polarization within the city. They may also spill out into the more remote urban fringe, where housing remains affordable. This relentless outward push, however, further increases the premium for desirable locations within the city, continuing the polarization of housing prices, which demands ever more capital inflows to sustain. Eventually, the bubble is punctured, and those caught in the "froth" face financial stress or insolvency.

The first of several boom-bust cycles in Southern California—in the 1880s, the 1900s, and the 1920s—were land booms. In McWilliams's (1973) classic account of these episodes, he emphasizes the "acute loneliness" (p. 168) that haunted the armies of newcomers, who came principally from other states in the United States. State societies, organized by these expatriates, proliferated in response to migrants' sense of social isolation. For example, the Iowa Society drew 150,000 to its annual picnics in the 1920s. McWilliams describes this federation as a promotion agency—giving information to tourists and home seekers, routing newcomers to the right real estate offices and banks, and so on.

Development as extensive and rapid as that of Southern California requires financing on a vast scale. Before the 1930s, this was usually provided by developers with large amounts of financial wealth and land rooted in silver and copper fortunes, the railroads, or land grant estates procured from the old California families. Since then, the primary source of finance for L.A.'s economic growth has been commercial banks and thrifts.

Financing the American (Suburban) Dream

The continued evolution of Los Angeles must be understood in the context of the New Deal legislation that rebuilt the U.S. banking system in the 1930s on the basis of two principles: market segmentation and government guarantees. Banking markets were separated geographically, and bank competition for savings and for loans was strictly limited. Government deposit insurance and limits on deposit rates ensured healthy bank profits. The Federal Reserve became the lender of last resort so that financial crises could no longer destabilize the financial system. In addition, savings and loan associations (thrifts) were established as a specialized sector for mobilizing household savings and financing home purchases. The invention of the 30-year mortgage and government mortgage guarantees put home purchases within the grasp of an increasing percentage of U.S. households.

The restoration of stable macroeconomic conditions and the lack of competing alternatives for savers and borrowers kept banks and thrifts profitable throughout most of the post-World War II era. In exchange for this regulatory protection, banks and thrifts provided broad-based access to deposit accounts, loans, and other banking services. All banking functions were local in the era before computerization, which meant that banks maintained branches in lower-income, as well as upper-income, communities.

The New Deal banking system provided a stake-holding vehicle for the broad mass of Americans. Most households stored liquid and savings balances in bank accounts and used long-term mortgages to purchase homes. Banks and thrifts coordinated the gathering and channeling of these savings, generating positive economic spillovers for the communities they served. Paralleling this financial stake-holding were new employment practices in American industry, which created many stable, high-wage jobs. Activist government policy was accepted as a means of managing macroeconomic growth. Together, these institutional realignments generated an economic boom. Sustained macroeconomic growth and robust industrial expansion encouraged urban and suburban growth, but highway subsidies, mortgage guarantee policies, and lower automobile prices ensured that the most rapid development occurred on cities' suburban fringes. Nowhere was this more true than in Los Angeles.

The benefits of this boom were not evenly shared, however. Unequal access to employment and financing channels left minority communities stranded in hollowed-out inner cities. Minorities and women were shunted aside in the labor market: They were displaced by returning soldiers after World War II and were disadvantaged by last-hired/first-fired rules in subsequent downturns. The banking system did not encourage wealth building within minority communities: Government mortgage guarantees explicitly discriminated against minority neighborhoods until the 1960s. Home ownership subsidies were of little use to most African Americans and Latinos, however, because of widespread residential segregation and racial earnings differentials.

Bank of America and the Taming of the Boom

Among commercial banks, the Bank of America (BA) has played an unparalleled role in the growth of Southern California. In building BA into the first U.S. statewide branch bank, its founder, A. P. Giannini, developed a strategy uniquely suited to a perpetual boom economy. Opening branches statewide allowed BA to mobilize the vast share of the disposable wealth flowing into the state. The resulting deposits were used, almost invariably, to finance the most pressing loan demands for industrial and agricultural expansion in the Golden State. In effect, BA was a stake-holding institution for Californians even before the New Deal reforms.

Bank of America survived the Depression years without incident. By 1954, it had 525 branches and 60% of California's deposits (Johnston, 1990, p. 21). This pool of reserves could be shifted as needed to meet the state's most profitable loan needs. BA used its hold over deposits to finance the Golden Gate Bridge, aqueducts and vineyards, the airspace industry in the 1940s and 1950s, and the electronics and computer industries in the 1960s (pp. 39, 62). The heterogeneous character of California's agriculture and industries ensured the bank a diversified loan portfolio (McWilliams, 1949, pp. 227-232).

The branches were BA's eyes and ears: All branch managers originated loans under the guidance of the bank's general loan committee. BA had no peer in coordinating the harvest of small savings, the finance of regional investment, and the capture of resulting benefits to further the regional accumulation of wealth. Although BA underwrote the state's more glamorous projects, it also competed with California's other banks and thrifts to capture deposits through branch networks and to provide the indispensable financing for residential development and small business.

The financing of the newly constructed single-family homes was primarily handled by thrifts up until the early 1980s. In 1965, Southern California was headquarters to only 3 of the largest 10 commercial banks in the western Federal Reserve district, but it contained head offices of 8 of the 10 largest thrifts and the second largest mortgage company in the United States, Western Mortgage. Thrifts grew rapidly in Southern California's residential boom conditions. Indeed, in 1965, thrift deposits in Los Angeles-Long Beach were larger than commercial bank deposits (McCann, 1966). Credit for small and intermediate firms came, for the most part, from commercial banks. For established (white-run) businesses, L.A.-headquartered banks such as Security Pacific Bank and Union Bank were important sources of financing.

Not all groups benefited equally from the New Deal banking structure. Prior to the Depression, people of color were informally excluded from many areas and sometimes removed when their presence was inconvenient. With the Depression, whites used minorities as economic scapegoats and created formal segregation barriers. By the 1940s, exclusionary zoning and restrictive covenants made 95% of Southern California off-limits to most minorities. The movement of minority residents into previously white areas has often resulted in reduced banking capacity and lending (disinvestment). The problem of disinvestment is evident as early as 1929, in this testimony of a resident of the University Addition neighborhood (just south of USC), into which Japanese and African American residents were then moving:

> To have Japanese and Negroes move in here would lower the value of our property. It is an economic problem much more than an ethical one. . . . The situation is rather difficult for us, too. If we want to raise money on our property, for example, the banks hesitate to loan as much on property where the colored people or the Japanese are encroaching. We neighbors must really stand together, for it is clear that if the man next door sold his house to a Negro or a Japanese my property would immediately become worth a thousand dollars less. (quoted in McClenahan, 1929, p. 85)

For minority communities, informal ethnic financial arrangements were crucial. Minority-run businesses often found access to credit through formal channels, such as banks, difficult or impossible (Dominguez, 1975). In the pre-Depression period, the extensive development of some minority communities—for example, East Los Angeles and Watts (Romo, 1983)—was accomplished by developers using bank financing. But more often—as in the case of the Mexican pueblo-plaza community—minority communities arose on the basis of their worker-residents' own slim resources. The paradigm case here is, perhaps, the Little Tokyo community near downtown Los Angeles. Little Tokyo grew up in an area previously occupied by boarding houses, whorehouses, and bars. The Japanese Americans financed their businesses, houses, and apartments through their own formal and informal financial institutions (Modell, 1977). These institutions survived and prospered because of these communities' cultural ties and also their social and financial isolation.

Surveys of the social character of Los Angeles throughout the 20th century indicate that areas such as Pacoima, Watts, and Boyle Heights have been populated disproportionately by minorities with lower-status jobs and lower incomes throughout most of L.A.'s history. A neighborhood's racial coding, its residents' economic status, and its place in the region's development schemes have all been closely interrelated. White areas were built and subsequently serviced by a robust banking infrastructure. Minority areas have historically lacked robust financial infrastructures and have been shaped by informal credit arrangements and anemic credit flows.

Global Capital and Global Competition

The U.S. banking system and L.A.'s financial dynamic have been altered fundamentally by a process of change that we term *globalization*. Beginning in the mid-1960s, macroeconomic pressures and growing sophistication by banks and their larger customers began to break down the New Deal banking system. Price inflation in the United States became more volatile. As a result, market interest rates often rose above deposit interest rate ceilings. In the 1960s, this led large savers to disintermediate—to flee the banking system for higher-yield investments elsewhere. In the 1970s, money market mutual funds emerged as a vehicle for smaller savers to disintermediate. As banks lost lending capacity, the large firms who were traditionally their most profitable and creditworthy loan customers circumvented banks by going directly to financial markets—first to the rapidly growing commercial paper market and later to the junk bond market.

In effect, the key features of New Deal banking—banks' monopoly over household savings and their primary role in credit provision—were being undermined. Bigger banks reacted by adopting aggressive new management philosophies: They sought out new loan customers—especially in less developed countries (LDCs)—to replace lost "blue chip" borrowers, and they replaced deposits with "purchased" funds such as certificates of deposit (CDs). Banks also restricted access in response to their profit squeeze: They closed branches in "marginal" communities, reduced services to smaller customers, and withdrew from some credit markets.

Beginning in the 1960s, the U.S. economy entered a prolonged crisis related to the globalization process: Corporations cut high-wage manufacturing jobs, sometimes relocating overseas, whereas lower-wage jobs grew, especially in the service sectors (Peterson & Vroman, 1992). This deindustrialization led to a "U-turn" in the U.S. income distribution (Bluestone & Harrison, 1988)—the relative growth of upper- and lower-income households, compared with the middle. The industrial decline, in turn, led to fiscal crisis and decline in many American cities. As the 1970s waned, many leading-edge U.S. banks began to shift their marketing strategies in response to these shifting fundamentals. Banks such as Citibank and the emerging superregionals[1] BancOne and NCNB designed retail banking services for upper-end customers while forcing lower-end customers to pay higher fees or dropping them entirely.

With the 1980s, bank and thrift financial distress rose as interest rates and the cost of purchased funds rose to unprecedented levels. Thrifts were especially compromised because of their specialization in long-term, fixed-rate mortgages. Reform acts in 1980 and 1982 deregulated deposit rates and granted thrifts new investment powers. A new government-backed secondary market allowed lenders to sell off their mortgages. Unfortunately, the dismantling of New Deal regulations worked no magic. Many thrifts used their new powers to support speculative projects and outright fraud. By 1989, federal legislation was again needed to bail out the thrifts' insolvent deposit insurance fund. Almost half the industry disappeared due to failures and takeovers. Meanwhile, the developing countries' debt crisis, which began when Mexico defaulted on its debt in August 1982, battered U.S. commercial banks. It took until 1987 for most U.S. lenders to take their losses and shed their nonperforming developing country loans. This restructuring only cleared the way for U.S. banks to plunge headlong into the late-1980s commercial real estate financing binge. When recession struck in 1990, the bursting commercial real estate bubble nearly forced a bailout of the commercial banking industry.

The restructuring of the U.S. financial system has largely overcome the geographic and product-line restrictions imposed in the past. Financial intermediaries once served customer bases that included the entire population of households and businesses within their market areas—with the exception of certain minority areas. These customer bases, however, have now been carved into at least three distinct market segments.

The most desirable segment is the *super-included:* the wealthy households that largely occupy the top 20% income areas within cities. These communities are well served by the formal banking sector. Indeed, despite an overall pattern of downsizing and branch closures, financial institutions continue to seek market share among these most desirable customers by locating new branches in these areas. Super-included customers have a wide range of options for conducting their financial transactions, meeting their payment obligations, and investing their wealth. Personal service and personalized products through brokers and account representatives are the norm. The intense competition for these highly profitable customers mandates this type of service.

Next are the *process-included:* the middle-income households who tend to live in the middle 40% income areas within cities. They too retain access to a wide range of financial services offered by formal institutions. They readily maintain bank accounts, accumulate personal savings, acquire houses and mortgages, and so on. They differ from super-included customers, however, because they do not receive personalized financial service. Their needs are met only because their financial characteristics are easily measured, standardized, and serviced. Financial services are provided as commodities, that is, impersonally and at low cost via ATM networks and computerized loan decisions through credit scoring.

At the bottom are the *process-excluded:* low-income households who tend to live in areas with the 40% lowest incomes. In the past, many of these households established relationships with banks and used these to overcome the barriers to creditworthiness posed by the difficulty of evaluating economic prospects in these areas or by weaknesses in household or business balance sheets. In the 1980s, however, formal financial institutions all but withdrew from the areas in which these households live. They have also priced their financial services in such a way that these households typically cannot afford to maintain accounts.

Financial Reinvention and the City of Angels

How have the changes associated with globalization affected the financial dynamic of Southern California? As noted, globalization began with disintermediation in the 1960s and 1970s. California's largest banks were most susceptible to disintermediation and the loss of large-firm, blue-chip borrowers, and hence were the first to feel the effects of globalization. The evolution of the Bank of America again offers an instructive barometer.

Bank of America reacted to blue-chip borrower loss by creating distinct lending offices for international and Fortune 500 customers in 1974. It aggressively made loans in developing countries; its lending volume trailed only that of Citibank. At the same time, BA reversed its financial services strategy and moved to consolidate its operations. BA's branch network peaked at 1,200 in 1980 when it held 35% of the California retail banking market. Beginning in 1980, BA closed branches and shifted loan-making decisions to centralized hubs. By the end of 1984, BA was down to 951 branches. Despite these adjustments, or perhaps because of them, BA neared insolvency in 1985 and 1986. Losses on its international lending, together with weakness in its California agricultural loans, forced BA to the edge of bankruptcy.

Bank of America's two major competitors in California reacted differently. Security Pacific Bank survived the worst of the LDC debt crisis and positioned itself as the "bank of the future." It rejected traditional information-intensive lending for sophisticated financial market transactions and services—aggressive positions on nonmortgage securitizations, foreign exchange operations, and emerging markets investments. This strategy proved fatal, however, and in 1991, Security Pacific was merged into BA. In direct contrast, Wells Fargo emphasized information-intensive, "back to basics" banking business. By the mid-1980s, Wells had acquired Crocker National Bank and was emphasizing retail banking while divesting its overseas assets and cutting costs. In the wake of its own difficulties, a humbled BA shifted course: It installed Richard Rosenberg, a former Wells executive, as CEO and sought to reposition itself in retail banking.

As the BA case illustrates, large lenders were forced to redefine their branches' functions. Large banks took loan originations away from their branches; loan decisions were instead made centrally on the basis of standardized information. Using their branches as sales outlets, the large lenders won over upscale retail customers by exploiting their dual advantages of scale economies and access to mass media. "Creaming" the retail market allowed the larger banks to draw more of their profits from fees and service charges, rather than from their traditional source, the ever thinner lending margin spread. In effect, larger lenders were mimicking the strategies and instruments of the nonbank lenders and financial market firms that constituted their stiffest upmarket competition.

For large California banks, the days of geographically matching the demands for credit and financial assets by households and firms were over. Smaller banks did not have the luxury of shedding geographic risk, however; their fate depended on the viability of the neighborhoods in which they lent. They were, at the same time, losing retail customers to larger institutions and nonbank competitors. Their survival depended on finding creditworthy borrowers in their service areas and then funding them at higher rates. Taken as a whole, bank behavior now accentuates, rather than levels, differences between the incomes and wealth of have and have-not areas. The informal financial firms that fill abandoned markets do not fully replace banks' services.

Just as BA began repositioning itself as a retail-oriented superregional bank, the very retail markets it was trying to tap were themselves being transformed. In the 1970s, Los Angeles had enjoyed disproportionate employment gains because of the balanced growth of manufacturing, producer and distributive services, and social service employment. In addition, Los Angeles did not experience spatial deconcentration between 1970 and 1980, and minority employment remained fairly healthy in these years (Johnson & Oliver, 1992). As the 1980s unfolded, deindustrialization and deconcentration occurred simultaneously in Los Angeles, with job losses concentrated in African American areas of the city (Johnson & Oliver, 1992; Soja, Morales, & Wolff, 1983). New growth *technopoles*—areas of job gains and plant openings—emerged in historically segregated suburban areas of Orange County and the San Fernando and San Gabriel Valleys (Davis, 1990; Scott, 1993).

As stable working-class jobs disappeared, lower-income residents of Los Angeles increasingly found themselves in full-time jobs at below-subsistence wages (Ong, 1989). Inner-core neighborhoods became more crowded, their political clout faded, and their public infrastructures deteriorated. Although income and wealth became more polarized, the absolute degree of racial segregation broke down. Legal and illegal immigration caused lower-income minority communities to spill over older ghetto boundaries. Legal challenges overturned racial covenants, giving minority middle classes more geographic mobility (Grigsby, 1994). Racial segregation has by no means been eliminated. Los Angeles, however, remains a hypersegregated city (Denton, 1994; Massey & Denton, 1993). Denton's hypersegregation analysis of Los Angeles finds that its 1990 African American population has, since 1980, become less centralized but more isolated and concentrated.

Continued in-migration at both ends of the income spectrum, together with the loss of well-paying manufacturing jobs, brought a U-turn in Southern California's income distribution. Manufacturing firms moved to the new exurban fringes, pulling traditional middle-class families with them. The vacuum within the city of Los Angeles was increasingly filled by new immigrants, primarily lower-income minority households. The dividends of the "slow growth" movement and Proposition 13 emerged as a decline in public services within the city. Wealthier households had two choices in the face of sprawl, overcrowding, and restructuring: They could withdraw into "privatized" space within the city—gated communities and private schools—or they could move to the exurban fringe with its standardized extensive development of the good life. Both responses served only to fuel the real estate boom in high-end areas while contributing to the hollowing out of the urban core.

In an earlier era, racial covenants and far-flung branch networks had helped to simplify banks' coordination problems, even while guaranteeing the physical decline of racial ghettos. Increased racial mobility, together with the consolidation of bank branches, now made it more difficult for banks to evaluate the "worth" of a loan for a house or small business within many communities across Los Angeles. The fluid and uneven character of postcovenant growth, together with leaner branch networks and centralized loan hubs, forced banks to rely on credit standards (scoring) that often shortchanged areas with new immigrants outside the formal banking sector. The values of properties and businesses in these areas became more uncertain—often because of the banks' and thrifts' own earlier racial lending patterns.

Booms, Busts, and Big Capital Revisited

Globalization and deregulation put immense pressure on California banks. Banks downsized staffs—California bank employment fell 21% between 1984 and 1991— closed branches, and consolidated operations (Federal Deposit Insurance Corporation, 1993). Many lenders failed. Banks that remained open sought new survival strategies. Many moved upmarket—seeking out wealthier customers and opening branches in new suburban and exurban upper-income communities. Among those moving upmarket were the large California banks.

This upmarket shift coincided with the Southern California real estate boom in the mid-1980s. This boom was generated by three factors. First, the region benefited disproportionately from the military spending buildup of the Reagan years. Second, the region was the target not only of a new round of domestic migrants but also of an unprecedented wave of immigrants from Latin America and Asia. Third, large volumes of overseas savings—both from immigrants and from current-account surplus countries on the Pacific Rim—inundated the region. As one indicator of the magnitude of this boom, single-family lending volume in Los Angeles averaged $6 billion annually from 1981 to 1983 but leaped to $39 billion annually for 1987 to 1989. Commercial banks plunged in to finance this boom—for the state as a whole, banks' residential real estate loans grew by 37.9% annually from 1987 to 1990.[2]

In this heady atmosphere, land and housing prices seemed to have no ceiling, and loans appeared to have little risk. Although many thrifts were sidelined because of insolvency, this boom was given a boost by the explosive growth of direct credit. Direct credit became far more significant in mortgage provision; the rise of secondary markets allowed brokers to originate and sell off mortgage loans without finding deposits to finance them. The real impact of direct credit, however, came in the arena of speculative land development per se. Previously, this development was coordinated centrally by larger landowners, large lenders, or both. Direct access to bond markets by many developers unleashed speculative purchases of land and the building of new cities. Land was speedily bought, parceled, and occupied on the exurban fringe—communities such as Moreno Valley and Lancaster. Fueled by competitive, decentralized, autonomously financed developers, this building boom quickly blossomed into widespread overbuilding.

Upmarket households were the main participants in the Southern California housing boom. Wells Fargo and BA led the way in moving upmarket in residential real estate lending. Wells Fargo's lending is consistently oriented toward the highest-income quintile. From 1981 to 1985, 53.9% of all Wells' residential real estate loan dollars went to homes in the 20% of census tracts with highest median incomes; by 1986 to 1989, 75.6% of their dollars were in these tracts. BA also shifted its lending pattern upmarket in the 1980s; by 1986 to 1989, 60% of BA's residential loan dollars flowed to the top 20% income census tracts (see Figure 4.1).

Figure 4.1 shows that in moving upmarket, the big banks were not simply following the overall market—they were choosing their own niche within the broader market. This

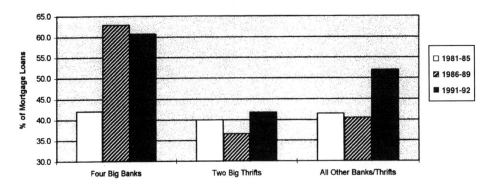

Figure 4.1. Bank Versus Thrift Lending in Highest 20% Income L.A. Census Tracts

NOTE: The four big banks include Bank of America, Security Pacific (merged with BA in 1991), Wells Fargo, and First Interstate. The two big thrifts are Great Western Savings and Home Savings. Data have been generated from a variety of sources.

figure contrasts the behavior of these big banks with that of the two giant Southern California thrifts, Home Savings and Great Western. These two thrifts make more residential loans than banks, and their lending volume dominates that of other thrifts, just as the big four commercial banks in aggregate lend far more than all other banks in Southern California. Figure 4.1 clearly shows that although the big banks' lending shifts during the 1980s toward the highest-income areas, the big thrifts' lending remains stable—no upmarket movement appears.

For lenders as a whole, residential loan volume in L.A.'s bottom 40% income areas remained well below the loan volume in the top 20% income areas. This bias toward wealthier areas is reflected in the character of the real estate boom itself. Figure 4.2 shows that average loan size grew much faster in the top 20% income census tracts than in middle- or lower-income areas. This "go-go" real estate boom, oriented toward upmarket customers and areas, took a variety of forms. On the one hand, extensive development led to the explosive growth of upper-income suburbs such as Chatsworth/Porter Ranch in the western San Fernando Valley; on the other, selective intensive development occurred with the gentrification of areas such as Santa Monica and Melrose.

Every boom comes to an end, and the Southern California real estate boom of the late 1980s was no exception. The preconditions of the boom were severed with the massive defense and aircraft cuts beginning in 1989, together with a slowdown in offshore saving and a national recession. Puncturing of the bubble had predictable consequences for banks. Many banks failed; although California banks' profit rates were spectacularly high from 1989 to 1991, they subsequently plunged. No soft landing was possible because beneath the visible regional boom lay the invisible decay of the communities left behind in the wake of the income U-turn and the banks' upmarket repositioning. Competitive pressures had forced commercial banks into narrow market niches, increasing the volatility of their returns and the volatility of economic growth within the city.

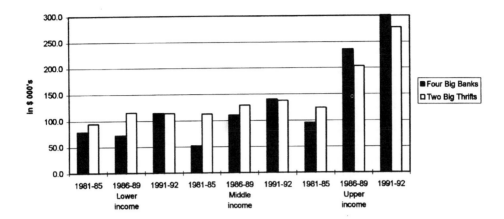

Figure 4.2. Average L.A. Loan Size Across Lender Types

NOTES: 1. Figures are for conventional single-family residential mortgages only, as reported under the Home Mortgage Disclosure Act. These data combine new purchases and mortgage refinancing. Data have been generated from a variety of sources.
2. "Lower Income" areas consist of the 40% of census tracts with the lowest average per capita incomes in the 1980 census tracts, "Middle Income" of the next 40% of the census tracts, and "Upper Income" of the 20% of the tracts with the highest average per capita incomes.
3. Average loan is derived by dividing the dollar volume of these loans in each category by the number of loans reported. This undercounts all averages because it ignores any secondary debt. Further, it systematically underestimates the average loan size in upper-income tracts because down payments there are typically a higher proportion of home value than elsewhere in the city. See Dymski and Veitch (1994).

The New Iowans: Pacific Rim Immigration and Capital

The processes of globalization that began in the 1960s have steadily increased the flow of financial capital across national borders. Have these trends made Los Angeles, as Soja (1989) predicted, a "capital of capital" for the Pacific Rim (p. 192)?

Soja's prediction of less than a decade ago has failed to materialize. The experience of California banks with overseas lending proved singularly unsuccessful. Security Pacific's aggressive strategy aimed at participating in global foreign exchange and securities markets failed in 1991, and it was taken over by Bank of America. Los Angeles remains, however, the home of some of the largest thrifts and mortgage companies in the United States—which only highlights Southern California's importance in the realm of consumer banking.

Although Los Angeles has not become a financial hub rivaling Tokyo or London, internationalization of capital has become a central part of L.A.'s financial dynamic in the past 15 years. Substantial capital flows from abroad have swept into the region. This pattern echoes L.A.'s earlier transformations from capital inflows, but although in earlier decades, these inflows originated from other regions in the United States, they have originated in other nations in the 1980s and 1990s.

Capital inflows to Los Angeles have taken a variety of forms, but they have involved a disproportionate investment in geographic assets, rather than financial assets. Most visible among these capital inflows has been the financing by Japanese, Hong Kong, and other Pacific Rim investors of large-scale, upmarket commercial development in Southern California. New commercial square footage in L.A.'s downtown and Mid-Wilshire business districts provide not only office space for Pacific Rim multinationals but also a source of portfolio diversification for investors in these dollar-surplus countries (Davis, 1987). These capital inflows contributed to speculative overbuilding in office space during the late 1980s, which exacerbated the real estate crash in the early 1990s.

Outside central business districts, however, an equally important wave of capital has taken place as a consequence of L.A.'s newest immigration pattern. Southern California's newest wave of immigration differs from past inflows by originating abroad, rather than from within the United States. In continuity with previous immigration patterns, however, some immigrants—especially those from Central America and Mexico—bring only their capacity for labor, whereas others—especially those from Asia—bring substantial amounts of both human and financial capital. Further, the scale of immigration has been so large that just as with African American and Latino in-migrations of earlier decades, the new immigrants have transformed L.A.'s residential and commercial spaces. Central American immigrants, typically arriving with little financial capital, have made Pico-Union a densely populated, low-income community. Sustained Asian immigration has overspilled the boundaries of older Asian communities, such as Chinatown, and led to the emergence of Asian "ethnoburbs" in suburban areas such as Alhambra and Monterey Park to the northeast of L.A.'s downtown (Fong, 1989; Li, 1995).

The combination of offshore financial investments in Los Angeles, the financial inflows of migrating households, and the accumulation of wealth by foreign-born households has led to the establishment within Southern California of a network of Asian- and Asian American-owned banks and thrifts within the past 15 years.[3] This financial nexus has helped fuel the rapid growth of Asian, and especially Korean, small business enterprises.

The blossoming of commercial and financial enterprise in these new Pacific Rim immigrant communities is itself a reflection of their social isolation. Many recent Asian immigrants, particularly Koreans, have substantial amounts of human capital (education and training) that they are unable to use in the United States because of cross-border differences in training, licensing procedures, language barriers, and discrimination. The availability of family labor at low cost, together with access to capital through ethnic banks or lending circles or personal savings, encourages these educated newcomers to become entrepreneurs. Often these businesses begin by serving clientele in areas dominated by African Americans and Latinos. As Asian ethnoburbs expand, however, these businesses increasingly serve their own ethnic communities, intensifying the social isolation of these new immigrant communities. Although the case of Korean businesses has received the most publicity, the pattern has been replicated for other immigrant groups as well. A tour of Pico-Union, for example, finds a plethora of Salvadoran, Nicaraguan, and Korean travel agencies, real estate brokers, doctors, and so on.

Although the overseas source of immigrants and capital is new, older patterns of adaptation are being re-created. The social isolation felt by Iowans and other newcomers from the Midwest in the early part of the 20th century led them to create state societies that developed into social and business networks. These institutions served to reduce the individual isolation and provided a framework for economic mobility—if not social integration. Similar institutions have been established today to provide support for many of the ethnic groups that have moved into the L.A. area.

In many ways, Asian immigrants, particularly the Koreans, are the Iowans of late 20th-century Los Angeles. The sustained commercial and residential development within Koreatown testifies to the power of ethnic financial institutions to channel the savings-lending process toward urban development. These ethnic banks have, by concentrating the savings and lending of their community members, re-created in a limited way older patterns of coordinated financial intermediation associated with Los Angeles during its suburbanization period. At the same time, the extreme financial hardship experienced by many Korean businesses in the wake of the 1992 L.A. uprising demonstrates the fragility of these small-scale development processes. The social isolation of new immigrants is intensified by their reliance on ethnic financial structures, which in turn leads to greater economic fragility for these new ethnoburbs.

Corner Store Finance: Nonbank Banking in Los Angeles

One outcome of the withdrawal of commercial banks from already established lower-income and higher-minority areas has been community disinvestment. The Mid-Crenshaw neighborhood once had a middle-class white population and a healthy financial infrastructure but was largely abandoned by lenders when its population turned African American. It has seen the growth of a secondary tier of financial service providers, distinct from the informal credit associations within immigrant communities. This secondary tier consists of market-oriented businesses, often small entrepreneurs, who substitute for the lack of bank branches in the area. The late 1980s saw an explosion of check-cashing outlets, pawnbrokers, and mortgage brokers who brought financial services to areas largely abandoned by banks and thrifts. Often these second-tier financial intermediaries are funded by the same commercial banks who withdrew from the area because it was "unprofitable."

Discarded by the formal banking sector, process-excluded communities turn to this growing second-tier financial sector to meet their financial needs. Check-cashing outlets and money orders are used for cash transactions and payments. These facilities are widely available to even the poorest households, for a price. Generally, the cost of these transactions varies inversely with the household's economic status. For obtaining credit, the options are fewer. Pawnbrokers have been described as the short-term credit market for the poor, but this is overly generous. Pawnbrokers act as lenders of last resort for the process-excluded. They offer credit on terms more onerous than those of the formal sector. Bank credit facilitates the accumulation of new human and physical assets that enhance future income; in contrast, pawnbroker credit involves households' deaccumulation of their stock of assets to meet current income crises.

Studies of informal credit markets focus on the cost of financial services as the key problem. Cost is sometimes, but not invariably, a problem in underserved areas. For example, check-cashing facilities often charge modest fees for steady customers. In addition, mortgage brokers originate much of the mortgage credit flows in underserved areas at market rates and terms. Other financial problems are more important for process-excluded communities. The main problem is these communities' lack of sufficient savings to sustain the levels of investment needed to reverse downward economic dynamics and to reignite growth. To some extent, this is the result of discrimination and of unequal access in labor markets. It is also, however, the legacy of uneven growth due to disparate credit flows in different urban communities.

From 1981 to 1993, at least 10% of bank branches in the city of Los Angeles were closed.[4] Two patterns appeared in this period: First, large numbers of early-1980s branch closures occurred throughout the area; second, this early-1980s branch downsizing disproportionately affected areas with large minority populations. Map 4.1 presents an overview of the vast changes in the financial service sector. The map divides Los Angeles into three categories on the basis of census tract median incomes: census tracts with the lowest 40% median incomes, census tracts with the middle 40% median incomes, and census tracts with the highest 20% incomes.

In the 40% lowest income areas, bank branches closed between 1981 and 1993 far outnumber the bank branches opened since 1988. The opposite is true for the highest 20% income areas. Bank branches that were in existence throughout the period (the black points) consist of a large percentage of branches whose ownership changed through merger or purchase during the period. Of the 817 branches counted in the 1988 Los Angeles branch population, only 40.8% remained open and with unchanged ownership status in 1994. Some 26.3% of these branches closed by 1994; the remainder were associated with an ownership or institutional change of some type.

Table 4.1 illustrates that these second-tier financial institutions primarily service lower-income communities in Los Angeles. The lowest 60% of census tracts in Los Angeles by income have the same number of bank branches for 1.5 times the population as the highest 40% income census tracts. This separation of financial services is presented in Map 4.2. Once again, census tracts in Los Angeles are divided into three categories based on census tract median incomes. Both Table 4.1 and Map 4.2 indicate that check-cashing outlets and pawnbrokers are overwhelmingly located in the lowest 40% income areas of Los Angeles. To a large extent, these lower-income areas are also high-minority areas. Hence, the withdrawal of traditional banks to upmarket locations has increasingly marginalized the ability of ethnic communities to benefit from the coordination of savings and lending that can strengthen their economic growth and wealth-holdings.

Table 4.2 illustrates the pattern of branch closure and openings during 1988 to 1994 for the lowest 20% income census tracts and the highest 20% income census tracts in Los Angeles. This table reports the median number of different types of bank and thrift branches and check-cashing outlets that fall within a 1-mile radius of bank branches that closed or opened between 1988 and 1993. Branches closed in high-income areas had more

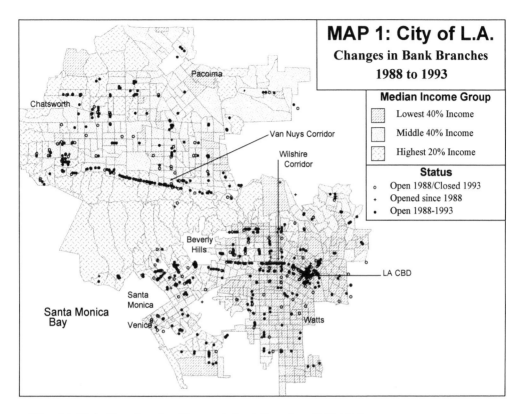

Map 4.1. City of Los Angeles: Changes in Bank Branches, 1988-1993

Table 4.1 Distribution of Financial Services in Los Angeles by Census Tract Median Income

1990 Median Income	Bank Branches	Credit Unions	Check Cashers	Pawn-brokers	1990 Population
Lowest 60% income census tracts	280	51	162	100	2,443,177
Highest 40% income census tracts	294	20	18	15	1,497,749

NOTE: These numbers are for 1993, generated from a variety of sources. The numbers exclude any bank branches in the Los Angeles Downtown or Wilshire Business Districts.

nearby branches than did branches closed in low-income areas, suggesting that these closures were simply responses to increased competition during the period. There were, however, a substantial presence of check-cashing stores near closed branches in the lowest-income census tracts. This suggests that although a significant demand for financial services remained, banks simply withdrew from low-income areas.

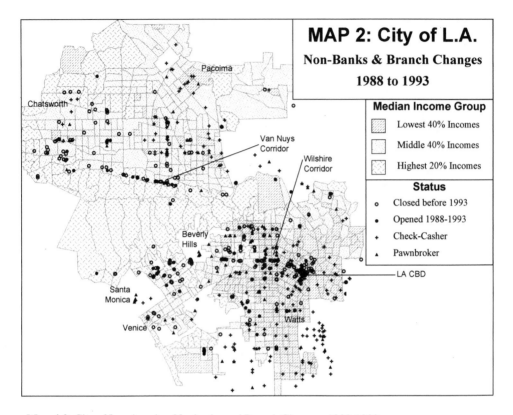

Map 4.2. City of Los Angeles: Nonbanks and Branch Changes, 1988-1993

Table 4.2 Density of Financial Services Around Bank Branches Closed and Opened, 1988-1993

Numbers represent the median number of financial firms of each type within a 1-mile radius of the bank branch closed during 1988 to 1993.

1990 Median Income	Bank Branches	Check Cashers	Total Branches Closed
Lowest 20% income census tracts	3	3	22 of 38 (58%)
Highest 20% income census tracts	4	0	55 of 189 (29%)

Numbers represent the median number of financial firms of each type within a 1-mile radius of the bank branch opened during 1988 to 1993.

1990 Median Income	Bank Branches	Check Cashers	Total Branches Opened
Lowest 20% income census tracts	17	1	9 of 38 (24%)
Highest 20% income census tracts	6	0	28 of 189 (15%)

NOTE: These numbers are for 1988 and 1993, generated from a variety of sources. The numbers exclude bank branches in the Los Angeles Downtown or Wilshire Business Districts.

Surprisingly, the same pattern holds for branches opened between 1988 and 1993: New branches are clustered in the highest-income census tracts near where other branches were already open; they are seldom near any second-tier financial firms. A comparison of the nearby-branch statistics for branch closures and openings shows, further, that branch openings occurred on average in areas already more densely covered with branches than did branch closures. Evidently, banks locate where their competitors already are or are moving to. Table 4.2 suggests a spatial separation between the formal banking sector—in ever hotter pursuit of upscale customers—and the second-tier financial sector.

Financial structures, as they have evolved to date, amplify growth and decay in urban neighborhoods. Second-tier financial firms that service lower-income neighborhoods do not provide a means of pooling savings to finance investment and the accumulation of new human and physical assets that enhance future income. These new firms meet the financial needs of a community in strictly limited ways. They charge higher fees for services and facilitate households' deaccumulation of their stock of assets to meet current income crises. In consequence, lower-income, high-minority communities are increasingly isolated from more prosperous communities. Social polarization results. This polarization encompasses not just the relative incomes available to residents of upper- and lower-income areas but also social well-being and health.

It All Comes Together, Separately, in Los Angeles

Soja (1989) suggested that "it" all comes together in Los Angeles—internationalization of production and labor, capital mobility, and spatial decentralization. We add to this that the people and capital that meet in this Los Angeles for reasons of fortune, exploitation, or even the heart, do so—to an increasing degree—separately.

Social and racial isolation have long existed in Los Angeles. The popular image of the city as a homogeneous whole was always a myth, attributable to a lack of contact among its many peoples in their different walks of life. As Jane Jacobs (1961) reported of an L.A. acquaintance in 1961, "Although she has lived in the city for ten years and knows it contains Mexicans, she has never laid eyes on a Mexican or an item of Mexican culture, much less ever exchanged words with a Mexican" (p. 72).

Los Angeles has normally generated this type of extreme separation because of the boom-bust character of its economic and spatial development. Extreme upward and downward shifts in land and housing values have continually created new classes of rich and poor. The new wealthy have bought their way in, whereas the poor have clustered in ghettos or been forced to the fringes.

The boom-bust character of the economy was tamed for a short time by Bank of America and the giant thrifts when regulatory restrictions gave them unchallenged dominance in coordinating the savings and lending process. They marshaled the tremendous savings pool to create the suburbanization of Los Angeles. When globalization and competition cracked open the segmented U.S. banking system, the period of coordinated urban and suburban growth disappeared. Global capital flows and increasingly direct access to markets for credit wrenched control of urban and suburban space from commercial banks and thrifts. In their stead, a myriad of smaller speculators, developers, and nonbank financial companies returned the financial dynamic of 1980s Los Angeles to the boom-bust character of earlier periods.

The previously unified banking markets also began to fragment in the late 1970s when faced with increased competition and new technologies. High-wealth and high-income households are now served by a wide variety of financial services provided by commercial banks. The middle class receives services from commercial banks and thrifts, to the extent that their characteristics are easily quantified and processed. For new middle-class immigrants, whose backgrounds and characteristics are more difficult to evaluate, new ethnic banks provide them with services that they would otherwise lack. In lower-wealth and lower-income areas, however, financially excluded households are forced to rely on informal credit to improve their economic standing. Market niche banking has resulted, and will continue to result, in inherently more unstable and disparate urban growth as increased competition within these narrower market niches reduces diversification and exacerbates the boom-bust character of real estate markets.

Notes

1. *Superregionals* are sizable banks with banking operations in more than one state but without global trading and lending operations. By contrast, *money center* banks are large banks that conduct banking and other financial operations in global financial centers.

2. Figures are derived from the Federal Reserve database for Home Mortgage Disclosure Act data.

3. Some banks, such as the Bank of Canton, have existed for a much longer period, as communities such as Chinatown and Little Tokyo have also.

4. We generated 1981 branch data from the Thomson Bank Directory for 1982. The data reported here undercount closures because this directory does not list all commercial banks and excludes thrifts. The data for 1988 and 1994 were generated from data provided by Atlas GIS.

References

Bluestone, B., & Harrison, B. (1988). *The great U-turn: The deindustrialization of America.* New York: Basic Books.

Davis, M. (1987). Chinatown, part two? The "internalization" of downtown Los Angeles. *New Left Review, 164,* 65-86.

Davis, M. (1990). *City of quartz: Excavating the future in Los Angeles.* London: Verso.

Denton, N. (1994). Are African Americans still hypersegregated? In R. D. Bullard, J. E. Grigsby III, & C. Lee (Eds.), *Residential apartheid: The American legacy* (pp. 49-81). Los Angeles: UCLA Center for Afro-American Studies.

Dominguez, J. R. (1975). *Capital flows in minority areas.* Lexington, MA: Lexington.

Dymski, G. A., & Veitch, J. M. (1994). Taking it to the bank: Credit, race, and income in Los Angeles. In R. D. Bullard, J. E. Grigsby III, & C. Lee (Eds.), *Residential apartheid: The American legacy* (pp. 150-179). Los Angeles: UCLA Center for Afro-American Studies.

Federal Deposit Insurance Corporation. (1993). *Historical statistics on banking, 1934-1991.* Washington, DC: Government Printing Office.

Fong, T. (1989, Summer). The unique convergence: Monterey Park. In M. Lopez-Garza (Ed.), Immigration and economic restructuring: The metamorphosis of Southern California. [Special issue]. *California Sociologist, 12*(2), 171-194.

Grigsby, J. E., III. (1994). African American mobility and residential quality in Los Angeles. In R. D. Bullard, J. E. Grigsby III, & C. Lee (Eds.), *Residential apartheid: The American legacy* (pp. 122-149). Los Angeles: UCLA Center for Afro-American Studies.

Jacobs, J. (1961). *The death and life of great American cities.* New York: Random House.

Johnson, J., & Oliver, M. (1992). Structural changes in the U.S. economy and black male joblessness: A reassessment. In G. E. Peterson & W. Vroman (Eds.), *Urban labor markets and job opportunity* (pp. 113-149). Washington, DC: Urban Institute Press.

Johnston, M. (1990). *Rollercoaster: The Bank of America and the future of American banking.* London: Ticknor & Fields.

Li, W. (1995). *Los Angeles' Chinese ethnoburb: Evolution of ethnic community and economy* [Mimeo]. University of Southern California, Department of Geography, Los Angeles.

Massey, D. S., & Denton, N. A. (1993). *American apartheid: Segregation and the making of the underclass.* Cambridge, MA: Harvard University Press.

McCann, W. (1966). *Los Angeles, financial center of the West.* Los Angeles: Department of Water and Power.

McClenahan, B. A. (1929). *The changing urban neighborhood: A sociological study.* Los Angeles: University of Southern California Semicentennial Publications.

McWilliams, C. (1949). *California: The great exception.* New York: A. A. Wyn.

McWilliams, C. (1973). *Southern California: An island on the land.* Salt Lake City, UT: Peregrine Smith Books.

Modell, J. (1977). *The economics and politics of racial accommodation: The Japanese of Los Angeles, 1900-1942.* Urbana: University of Illinois Press.

Ong, P. (1989). *The widening divide.* Los Angeles: University of California, Research Group on the Los Angeles Economy.

Peterson, G. E., & Vroman, W. (1992). Urban labor markets and economic opportunity. In G. E. Peterson & W. Vroman, *Urban labor markets and job opportunity* (pp. 1-29). Washington, DC: Urban Institute Press.

Romo, R. (1983). *East Los Angeles: History of a barrio.* Austin: University of Texas Press.

Scott, A. J. (1993). *Technopolis: High-technology industry and regional development in Southern California.* Berkeley: University of California Press.

Soja, E. (1989). *Postmodern geographies.* London: Verso.

Soja, E., Morales, R., & Wolff, G. (1983). Urban restructuring: An analysis of social and spatial change in Los Angeles. *Economic Geography, 58*(2), 221-235.

Thomson Financial Publishers. (1982). *Thomson bank directory.* Skokie, IL: Author.

Accountability and Responsiveness

ABBY J. LEIBMAN
Executive Director, California Women's Law Center

I have tremendous concerns about the future of the city of Los Angeles and the influence of the mayor's office on this future. I'm worried about the lack of access for some portions of the community. I'm worried about lack of accountability by and of our elected officials. I am concerned because I do not believe that a corporate model can be translated into government. It seems to me that there are central differences between what it means to be in government and what it means to work in a private corporation. The biggest differences have to do with issues of accountability and responsiveness. They work for us. I'm worried that this may not be true anymore. I also realize that the sharing of power is a difficult concept. When you've got it, it's hard to let go of it. I intend to press so that power is shared with those people who have been excluded from it for far too long.

Is Los Angeles Governable? **5**
Revisiting the City Charter

H. ERIC SCHOCKMAN

Public Enemy No. 1 in Los Angeles is not any character under scrutiny of the grand jury. It is the Los Angeles City Charter—a 125,000-word legal spider web of entangling phrases in which any honest official, once caught, struggles vainly for release. The only beneficiaries of this web are the nameless spider architects who have a vested interest in chaos, conflict and controversy.

Ransom M. Callicott (1949),
former commissioner, Board of Civil Service,
and former city councilman (from 1955 to 1962)

Historical Perspective of the City Charter

The city charter is the bible for governance in the city of Los Angeles. As the basic law of the municipal corporation of the city, it identifies the officers, as well as their powers and duties; defines the functional operations of the various city departments and commissions; defines the general provisions for the conduct of municipal elections; and concretely spells out the finance and budgetary operations in the city. The original document, adopted in 1925 (*Charter of the City of Los Angeles,* 1925/1973) is a massive work, comprising more than 400 pages and 500 sections that spell out the 34 articles of incorporation.

Under Article XI, Section 8 of the California Constitution, any California city containing more than 3,500 people may adopt its own charter by setting forth the procedures outlined in this section. Specifically, the process is initiated by holding an election to question the drafting of a charter and the selection of a board of 15 freeholders. If a majority of voters favor drafting a charter, the top 15 freeholder vote-getters in the same election are given the charge to draft a charter within 1 year. One built-in option to the freeholder drafting method is to have the charter framed by the city council or its representatives.

On completion of the charter draft, it is submitted to a majority vote for ratification. The local charter must be presented to the California Legislature in its next sitting session and will become ratified only if it receives a majority of votes in both houses. After legislative approval, it is then deposited and filed with the secretary of state.

The legacy of freeholder charters demonstrated the shifting of political power away from absolute state control to the doctrine of "home rule" for cities. Since the founding of Los Angeles on September 4, 1781, as *El Pueblo de Nuestra Señora la Reina de los Ángeles de Porciúncula,* the supremacy of power for the governance of the city resided with the state. From 1850, when Los Angeles was incorporated as the first city in the new state of California, until 1889, when the city's first locally drafted charter went into effect, power remained in the hands of the state legislature.

The Charter of 1878 (based on the amended Incorporation Act of the State of California) not only made substantial alternations in power relations between the state and the city but simultaneously altered the *intra*city balance of power. The charter set up a council of 15 elected members by districts, elected for 2-year staggered terms. It further strengthened the power of the mayor by mandating a four-fifths override by the council to sustain a veto and permitting the mayor to have membership on the council's standing committees. Moreover, it continued to support the concept of citizen boards and commissions as a ward of mayoral control and power. This period was probably the historic zenith in the development of mayoral dominance and control in the evolution of Los Angeles.

On May 31, 1888, a board of 15 freeholders was elected by the voters of Los Angeles to prepare the first local charter for the city. By a vote of 2,642 to 1,890, the citizens approved the Charter of 1889, which did two important things for Los Angeles: (a) It *philosophically* cemented the doctrine of home rule as a natural order of democratization of municipal affairs, leaving power in the hands of local citizens, not legislators; and (b) it *structurally* anointed the city council as the most powerful component of government, shifting dominance away from the mayor by controlling such things as the power of the purse, a lesser threshold to override mayoral vetoes, and confirmation of certain mayoral appointments. As John C. Bollens (1963) reminds us, there were five major attempts to obtain a new charter in the 36-year period between 1889 and 1925, but only the last one was successful (p. 45).

The significance here appears to be the solidification through time of the trust and belief of the voters of Los Angeles, *vox populi,* in a charter of city government based on a weak mayor-council-commissioner-administrator form of government. This early cumbersome foundational structure is often cited by scholars on local government as an extreme example of separation and diffusion of authority and accountability (see for example, Banfield & Wilson, 1963; Connery & Caraley, 1969; Lowi, 1964). What began in the Charter of 1889 and was codified in the Charter of 1925 (the current operating *Charter of the City of Los Angeles,* 1925/1973) was the groundswell of revulsion with political machines and corruption of urban East Coast regimes. Add to this the domination held over California politics by the giant railroad companies, and the result was the onset of the reform movement and the Progressive Era in California.

Before discussing the profound impact and details of the Charter of 1925, let me just deviate for a moment to discuss the "good/bad/ugly" aspects of the reform movement and its legacy on contemporary governance, the city charter, and community bonding in Los Angeles. The dictum I wish to apply here goes like this: "What was once bad is now good, or that which is currently perceived as good is inherently bad." The Progressives intentionally sold honest government as good government. "Yet honest government is not necessarily good government but simply its prerequisite" (*Report of the Los Angeles City Charter Commission,* 1969, p. 190).

The "good" of the reform movement produced a spiderweb of extensive dispersion of governmental power—very much a product of the philosophical minds of the late 19th- and early 20th-century political leaders. Fragmentation of power in municipal government was supposed to provide cogent checks and balances against runaway machine politics. Even more organic was the fundamental belief that centralized power is inherently evil, which meant that if the choice was centralized democratic control or a rigid maze of impediments to promote "honest" (yet sometimes ineffectual) government—the choice for the Progressives would fall to the latter.

Los Angeles was the leader in "direct democracy" legislation among established cities of the time, and even years ahead of state voters who later ascribed into the California Constitution the direct democracy measures of the initiative, recall, and the referendum (see Allswang, 1991; Deverell & Sitton, 1994; Key & Crouch, 1939; Mowry, 1951). In 1902, Dr. John R. Haynes led a Progressive coalition in orchestrating the passage of these measures as charter amendments in Los Angeles.

The quality of the "good" of these measures from a historical perspective needs to be critically assessed. Did direct democracy bring cleaner, more honest government from 1902 to 1903 on? Absolutely not. During the 1930s under the administration of Mayor Frank Shaw, the city was one of the most corrupt municipalities—on a comparative mode with the vice, gambling, and corruption of Chicago. The election of 1938 brought in Judge Fletcher Bowron to reclaim the reins of city governance and restore ethical trust. The members of the Los Angeles City Charter Commission of 1969, in discussing those historic reform measures embedded in the 1925 charter, wrote,

> In some respects the detailed and confusing nature of the Charter might actually prevent corruption from being easily detected. The diffusion of authority in the Charter makes it difficult to trace a problem to its source, facilitates disclaimers of responsibility by city officials, and prevents the clear placement of blame for deficiencies. (*Report of the Los Angeles City Charter Commission,* 1969, p. 190)

Although the referendum and initiative process started out to break the logjam special interest held on the legislative, policy branch of government, it has come full circle. Today, it symbolizes the diminution of elected leadership and the reentrenchment of special interests who use the ballot box to avoid deliberative debate. Combined with the structural heritage of the city charter, these "reform" measures have ironically contributed to the maldistribution of equitable power and resources among all citizens of the city.

Other "goodly" intended measures of the Progressives, in my opinion, have reinforced today the gap between municipal government and the city's populace *writ large*. In 1909, for example, the Progressives were able to convince Los Angeles voters to adopt a charter amendment that provided for nonpartisan elections. At the attempt of weakening political parties, it meant that *public relations, not party discipline,* would become the key to electoral and, ultimately, governing success. Political parties have traditionally served as ideological mass mobilizers for the electorate, as well as mechanisms for grooming, screening, and controlling future leaders. Politics as "personalized," has been replaced by "politics of media enhancement." The flow of political information and the Jeffersonian faith in the common rational yeoman is in control of those who can purchase and package it.

Not only has this become an electoral sham, "no-party politics" described in the city charter makes a travesty of coherent public policies. Because the charter gives the council supremacy in governance and policy making, it would be this body that would be most assisted by the advent of party loyalty, responsive to party leadership and discipline. Yet because the Progressives sold us the mythology that "there is no Republican nor Democratic nor Independent way to fill a pothole," we see alliances in the council as transitory phenomena, devoid of long-term ideological vision ("the vision-thing") and based solely on the flux of individual issues as they appear on the political landscape. Policy has suffered proportionally, and as observers have seen, getting consensus out of the council is "like herding 15 cats to go in one direction."

Furthermore, present-day multiethnic, polycultural Los Angeles lacks the mediating institutions across socioeconomic communities that are needed to sustain consensus building and coalition politics. Political parties would have provided that cross-fertilizing mechanism had not the charter precluded them from such a critical role in municipal governing (see Schockman, 1990, p. M5).

These are but a few of the "good" legacies of the reform era enshrined in the charter that have soured through the decades and produced dysfunctional government in Los Angeles. Perhaps there are scholars who can explain away this good/bad paradigm, dismissing them as harmless vestiges of a time-bound period in the evolution of urban government in America. Perhaps they will offer the rationale of why they are still with us, that is, we haven't been able to reinvent anything new and solid to replace them with. Perhaps this naïveté even has some merit.

If one can explain away the good/bad paradigm, one cannot explain away the "ugly," more diabolical analysis behind the reform movement in the development of urban American. The ugly side of the reform movement in general, and the case of the Progressives in Los Angeles specifically, was that it was a *class and cultural war* between the newly empowered immigrant/ethnic classes versus the rule by the educated upper and middle class. The ugly dimension of this was that the class bias of municipal reform was sold in simple terms as a business model of urban government efficiency. Reformers were neither apolitical nor unaware of class antagonisms. This was non-Marxist class warfare concealed in the city charter. As Samuel P. Hays (1992) notes, "behind the debate over the method of representation, lay a debate over who should be represented, over whose views of public policy should prevail" (p. 219).

This intentional ugly side of the reformers was much more than the musings of muckrakers, such as Lincoln Steffens, Theodore Dreiser, and Upton Sinclair, against corruption in the nation's big cities. It was the formation of a national blueprint for the business and upper-class elites to close ranks against the onslaught of immigrant constituencies and frame the debate as to home rule charters, electoral reform, and new agenda setting. It was, furthermore, a profound *cultural* clash between what Banfield and Wilson (1963) identify as those who were "private regarding" and those who were "public regarding." To the immigrant, new ethnic communities, all politics was personal and had a *private ethos*—whether that was a payback for a vote or a patronage appointment. To the reformers, the idea of the *public ethos,* of efficient government, seemed appealing because it promised social betterment without a change in class relationships. It was to become the "Quiet Revolution," a panacea to end ethnic strife, worker discontent, and municipal agitation. Of course, the irony today is that there has been a complete reversal in roles between those who are public regarding and private regarding on the municipal sphere. Urban political orders today controlled by the "haves" are much more prone to be private regarding. In Los Angeles, the "capital of the tax revolt," which produced Proposition 13 in 1978, and whose privileged Westside and San Fernando Valley communities today clamor for NIMBYism and gated and walled communities, we witness the end of reformism whose true manifestation has become the private ethos. Leave it to the "have-nots" of contemporary Los Angeles—the newly arrived immigrant classes or the long-term ethnic permanent urban underclass—to clamor for the public-regarding resources such as quality public schools and decent health care (see Judd & Swanstrom, 1994, especially chap. 4; Schockman, 1991).

This abbreviated historical perspective spells out the amazing resilience and adaptation of the city charter to inscribe castelike power configurations to the governing of Los Angeles well into the twilight of the 20th century. It speaks to the very essence of the rigidity of incorporation of contemporary inner-city constituencies and to the peripheralization of non-Anglo, non-middle-class communities who happen to be the plurality of the city's residents. It demonstrates in part the dysfunctional governing coalition that artificially administers the metropolis and that by mere necessity of this vestige position, has kept clear of implementing charter reform and its resultant possibility of a shift in the balance of municipal power.

The Charter of 1925 and Charter Reform:
The Convergence of the Good, the Bad, and the Ugly

The 1925 Charter is truly a masterpiece of its historical origins and a brilliant reform document of its time. Of the 30 candidates who sought to be elected to the board of freeholders in 1923, 15 were chosen to draft a new charter. On May 6, 1924, the voters of Los Angeles passed Proposition 1 by a margin of 126,058 to 19,287. It ushered in the form of government that for almost 71 years now had played havoc not only on the functional maturity of administrating city government but also on thwarting fuller political incorporation by not awarding stake-holding privilege to those most disenfranchised in the political process.

Charters are typically conceived as instruments of the people to direct the form of their city government. Charters should be concise, timeless, generally consistent, clear, flexible, and responsive to people. Charters should be social contracts between the municipal rulers and those who are ruled, and although their main purpose is to provide restrictions on government, their powers should not divest communities in improving access and participation in all aspects of municipal affairs. Unique about the 1925 Charter is the duality of its coloration: clean government with dispersion of political power and the simultaneous creation of what I term *municipal feudalism*—in which barons and plebeians play scripted roles in assigned caste power arrangements.

If the city is a democratic institution and the charter its oracle for governing, the 1925 Charter stands in stark contrast in translating the needs of diverse neighborhoods and communities into meaningful and vital political ends. The current charter was designed when Los Angeles was *the most* white, Anglo-Saxon, Protestant city in the nation. Today it is one of *the most* ethnically diverse communities in the world. The focus on the structure of governance as personified by the charter is critical in seeking solutions on how racial, class, and ethnic competition will be resolved in the future.

The External View: The Charter Versus Neighborhoods

In examining the 1925 Charter from *outside* the mechanical operations of city hall, one might return to the concept I previously advanced: the creation of a municipal feudalistic structure. Communities of interest and neighborhood power bases were systemically dismantled under the charter through the years to prevent alternative power structures from threatening the 15 "council-barons" (or 15 minimayors, as some would claim). Channeling and controlling increased citizen and community participation through the councilmanic offices became ends in themselves. With the return from at-large to district elections in Los Angeles, communities were played off each other, simply bought off in the 1970s and 1980s with federal largess money, or reapportioned every decade to prevent consolidation and political mobilization. The late councilman from the "Great 9th," Gilbert Lindsay, used to refer to himself as the "Emperor of the 9th Council District." Self-proclaimed titles such as this are important in understanding the complexities of this municipal feudalistic arrangement. "Emperor" Lindsay was *the* key to unlocking the fortunes of the downtown redevelopment plan and had to form an unholy alliance with the mayor, the Community Redevelopment Agency, the downtown developers, and the business community to take care of affairs in his district. Unfortunately, it was to the detriment of the mostly poor, African American, and Latino communities who inhabited the rest of the "Great 9th." With weak balkanized communities developed intentionally by the charter's structure, little of the economic goodies that were the side products of the downtown building boom ever reached the peripheral residential, mostly minority communities.

Communities and neighborhoods within the city have always existed in some embryonic form. The sad history of charter reform in Los Angeles is that it has usually treated neighborhood empowerment as a sideshow to structural overhaul for administrative efficiency. The general nature of the problem, however, will not just go away and is demonstrated continuously by social movements in L.A.'s neighborhood history to either secede from the city or to establish boroughs to obtain a degree of self-determination.

Throughout the chronological periods to define community empowerment in Los Angeles, these movements have been orchestrated not by ethnic or racial communities of East or South Central[1] Los Angeles, but by the wealthy sections of the Harbor and the San Fernando Valley. Back in 1909, the residents of the cities of San Pedro and Wilmington—who were voting on the mandate of consolidation with Los Angeles—inserted into the charter a key provision for the creation of boroughs to retain a degree of self-government and community. Although invalidated by a California Supreme Court of 1917, the concept of a "community of boroughs" held out as a romantic notion of returning city government to its community origins. Later, discontent in the San Fernando Valley brought on years of debate around secession from the city. To a large extent, the debate was about city resource allocation north of the Santa Monica Mountains and the heavy tax burden "valleyites" pay proportionally into the general fund. To a lesser extent, the subdebate was about the "retention of neighborhoods" (which up until the 1960s was largely white and middle class). Secession would bring local control back for keeping neighborhoods intact—having decisional control over zoning, housing, parking, and transportation. The city hall fathers (as sexist as reality was) saw the handwriting on the wall early on and took the wind out of the sails by tinkering with government reorganization by providing for *administrative* and *territorial* decentralization, whereas holding back on *political* decentralization (Costikyan & Lehman, 1972). Thus, for example, the opening up of a Valley City Hall and decentralizing the building and safety department was enough tokenism to gut the considerable tide in favor of boroughs and secession. Neighborhoods and communities, *stricto sensu,* were the ultimate losers in this battle.

The tensions apparent between the charter and the neighborhoods in Los Angeles in 1992 (with the outbreak of civic unrest) was not substantially different than it was in 1965. The National Advisory Commission on Civil Disorders (the Kerner Commission) (1968) found that

> city government appears distant and unconcerned, the possibility of effective change remote. As a result, tension rises perceptibly; the explosion comes as the climax to a progression of tension-generating incidents. To the city administration, unaware of this growing tension or unable to respond effectively to it, the outbreak comes as a shock. (p. 288)

It was evident that members of the 1969 Los Angeles City Charter Commission, appointed by Mayor Yorty in a 28-month comprehensive review of the 1925 Charter, heard and digested the Kerner Commission's recommendations. To their credit, they treated, perhaps in the most serious manner to date, the implications of the charter in recognizing neighborhood government and promoting community representativeness. They added to the proposed "Charter Solution" (a) a self-defined neighborhood organization plan, (b) an elected neighborhood board, (c) an appointed neighborman, and (d) the creation of a city ombudsman office. Although this particular charter reform effort failed, as so many previous efforts did, the work of the citizen body still carries a salient message. In their concluding comment on dispersing power *away from* city hall and back to the neighborhoods, their righteous vision speaks as clear today as in 1969:

> Los Angeles has one of the largest city populations and areas in the world, and the meaningful involvement of its citizenry is a basic and great challenge. The commission believes that making it possible for local residents to create this new type of public organization at the neighborhood level as part of the city government will significantly alleviate alienation and increase the level of participation. Of course, this mechanism will not solve all the human and social problems. It will, however, open new and greatly needed avenues for communication and participation. (*Report of the Los Angeles City Charter Commission*, 1969, p. 27)

Summarizing this external view of the impact of the city charter provides a sense of a deliberate calcification around the status quo power arrangements that have perpetuated this functional municipal feudalistic structure in the governance of Los Angeles. This has come partially through the centralizing of political power and administrative authority within the confines of city hall—at the expense of neighborhood or community empowerment. Any radical surgery on charter reform today must include this key external variable to address citizen participation and neighborhood advancement, which would break the lock on municipal feudalism.

The Internal View: The Charter as a Flawed Document

In reviewing the literature on the city charter, one seminal study conducted for Town Hall by Bollens (1963) stands as the single most comprehensive analysis of the dissection of the 1925 Charter. From the perspective of *inside* the mechanism of city administration, Bollens's analysis not only critically scans the decades of the council's resistance to take action to fix the inadequacies of the 1925 Charter but also synthesizes the most consistent recommendations for change, which are these:

1. Increase the administrative authority of the mayor (or occasionally the chief administrative officer), so that "he can be justifiably held accountable for his charter responsibilities."
2. Make boards and commissions advisory, eliminate them, or possibly retain them for appeals purposes.
3. Provide the mayor with appointive and removal authority over department heads and general managers (sometimes the recommendations call for approval of the council, sometimes not).

4. Relieve the city council of administrative authority; it should be concerned with legislative policy functions.
5. Reduce the number of days of the week the council meets, elect some at-large members, and enlarge or reduce the size of the council.
6. Consolidate the departments into a smaller number that is more focused on current needs.
7. Remove the administrative detail of the current charter and put it into an administrative law code that can be changed by ordinance without going to the voters. (pp. 40-42)

A Citizens Ad Hoc Working Group for Charter Reform in 1993 added on to and essentially updated Bollens's survey to include the following needed internal reforms:

1. In the absence of mediating institutions, there is a need to develop a structure that encourages coalition and consensus building, possibly through the political parties, possibly some other mechanisms.
2. The role of city government in the city's future should provide for a supportive infrastructure for economic stability and growth.
3. The complexity of decision making makes reaching a conclusion a Herculean task. Authority is dispersed too widely, given the involvement of county, regional, and state agencies, but at least within the city, an attempt should be made for some reformulation of agencies to simplify the path.
4. Some mechanism is needed to respond to changing needs. Needed is a balance between stability and responsiveness to rapidly changing circumstances. This might also involve rethinking job descriptions and requirements for promotion, assuming civil service is retained.

From this laundry list of policy recommendations for charter revision, I wish to focus on just one aspect—the functions of boards and commissions under the current system—as illustrative of power distribution and policy outcomes in Los Angeles and, I venture to say, a fortiori, most of urban America. An analysis of the composition of boards and commissions may lead to a more cogent perspective of the political accommodations the present charter awards. As urban political scientist Clarence N. Stone (1989) uncovers in most of his research, what makes governance work in any city or "urban regime" is not the formal machinery of government but the *informal* arrangements and partnership through which major policy decisions are made. These extraformal arrangements, therefore, need to be examined beyond the conventional analysis of structural outputs of the institutional parts of city government. The legal agreements and formal charge of the city charter, which are kept out in public and press view, have been the best disinfectant against blatant malfeasance and corruption. The present city charter breaks its covenant with the people in many arenas; it is the formulation and empowerment of boards and commissions, however, that limit the agenda narrative to the result of what the governing regime wants. Policy outcomes are manipulated behind the scenes in the information partnerships that commissioners share with their nominating mayor (and the mayor's allies on city council). This has produced a municipal feudalistic dominion in the governance of Los Angeles, as commissioners play the role of loyal feudal knights operating in semiobscurity and carrying out the commands of the governing urban feudal regime. The vehicle that has made all this possible is the standing city charter.

Boards and Commissions: The Shadow Government

The lay board and commission system in which the city of Los Angeles has incorporated since 1869 is a labyrinth of bizarre power arrangements between the citizens, the mayor, the council, and the city bureaucracy. Under Section 78 of the charter, direction is given so that the

> head of each department shall have power . . . to supervise, control, regulate and manage the department and to make and enforce all necessary and desirable rules and regulations thereof and for the exercise of the powers conferred upon the department by this Charter. (*Charter of the City of Los Angeles,* 1925/1973, p. 61)

A great amount of subtle and overt power is dispersed to these boards and commissions under the charter: the power to appoint and remove general managers, the power of administrative policy setting for each department, the power of full charge and control of all work of the department, the power of overseeing budgetary appropriations and expenditures, and the power to control the flow of public policy outcomes.

Politically, not all boards and commissions share equal power nor command and respect. The Charter of 1925 did standardize and harmonize certain aspects of their composition—for example, boards and commissions shall consist of five members,[2] appointed for 4- or 5-year staggered terms by the mayor and removable by the mayor,[3] subject in both cases to council consent and, under current policy of the Riordan administration, are not to receive any compensation for their services because of the severe budgetary crisis in the city. The charter gives specific authority to three departments (boards)—airport, harbor, and water and power—to act as independent, self-governing bodies of the city outside the general fund because they are revenue producing and proprietary municipal corporations and, hence, should be exempt. Enormous power is tied to each of these "proprietary" entities and in truth speaks to the unfettered (some would claim unaccountable) economic past and future of the L.A. basin. Water and power made growth possible and arid land bloom. Although the Municipal Water Department was the recipient of a $24.5-million public debt to contract a 223-mile water pipeline to the Owens Valley in 1905, key city leaders who were also large San Fernando Valley property owners got richer as the water brought increased population and growth. William Mulholland, Harrison Gray Otis, and Harry Chandler were but three whose entrepreneurial and visionary skills established early linkages in the privatization of the public domain (Deverell & Sitton, 1994; Mowry, 1951; Ries & Kirlin, 1971). No less bold was the movement of Los Angeles to the Pacific Ocean and the promotion of a $40-million bonded indebtedness to secure a harbor at the port of San Pedro. Today, the Los Angeles Harbor is one of the busiest in the nation, and the Los Angeles Airport nationally ranks fifth in the amount of air freight cargo it handles. With the signing of the North American Free Trade Agreement, Los Angeles and its key public facilities are in an advantageous geographic and financial position to prosper accordingly (see Schockman & Madjd-Sadjadi, 1994).

In the history of Los Angeles, the synergy between public improvements and power can be traced through the evolution of the charter's "shadow government" created through its boards and commissions. By connecting the flow of public money to dots of the historical picture, one can trace the power of commission appointments through the domination of the early 1900s of the Department of Water and Power, through the growth boom of the 1950s to 1980s and the Planning Commission and the Board of Public Works, and in the 1990s to 2000s into the $60 billion of public funds for transportation infrastructural development controlled by the Metropolitan Transit Authority (MTA) and the Transportation Commission.

In current reality, power has shifted as money has shifted away from the heady days of massive infrastructure development to transportation growth. The MTA board is where the action is for the remainder of the decade. If one assumes, then, that power is never stagnant and power relationships are fluid, why haven't the citizens of Los Angeles challenged the theoretical assumption for the retention of the present board system, short of overhauling the entire charter?

Let me return to the general discussion of boards and commissions as they present themselves today in the charter. Some great mythology and political naïveté have established themselves around the structure of these citizen boards and the conviction that boards were a means of promoting citizen participation in government. The rationale for boards and commissions lies in some 19th-century, agrarian, homogeneous society, ill-suited for postmodern, posturban, polycultural societies. The arguments in *support* of continuance of such a system have such wholesome democratic pluralistic overtones. For example, supporters believe lay participation provides community representation and a wide range of knowledge from diverse fields of specialization, which can in some way balance the dictates of the professional general manager.[4] The board system, in theory, is the intermediary between the citizen and local government. Board members help steer the general managers and the departments to hear, see, and listen to the public on matters of public policy. The board system preserves the checks and balances needed in urban democratic institutions. Authority is decentralized and prevents corruption from taking root. Board members furthermore are "independent thinkers" representing the diversity of the city's rich, complex social and class fabric—not just a rubber stamp for their general managers or for the mayors that appointed them.

A dose of political truism usually can debunk this mythological philosophizing, especially in the case of uncovering raw naked power through the tools of the city boards and commissions. A "shadow government" has been created under the Los Angeles city charter, defusing power on the surface yet centralizing power to the detriment of the average citizen. Far from being nice offshoots of the postreform era, the city boards and commissions determine the policy agenda of their respective agencies and determine, in the Robert Dahlian sense, "who gets what, when, and how."

This general critique argues that part-time (lay), amateur boards cannot possibly administer to the complexities of postmodern Los Angeles. No matter how proficient outside members may become in the functioning of departments, they essentially become "wards" of the expert knowledge base held by general managers and the city staff. This last point became evident in the handling of the post-Rodney King incident within the Police Commission. It took the ad hoc Christopher Commission and the passage of Charter Amendment F to place into operations a more independent source than the Los Angeles Police Department to give informed data to the members of the Police Commission so that they might objectively determine policy on the gamut of police matters. Furthermore, the entire system of boards and commissions is a contraefficient model of operating complex units of city government. The charter has stultified interdepartmental coordination by organizing each unit *vertically*—with its own board and internal politics. Efficient management is not by group dynamics but by centralized authority in the form of the general manager, who would ideally be accountable to the citizens via the mayor (or under present conditions, via the majority of the city council).

A more specific *political* critique is duly in order. Let's pursue this line of questioning: Who exactly are these shadow commissioners? Where do they live? Whom do they represent? To whose loyalties do they pay homage? Do they reflect the diversity of Angelenos and deserve to serve on a "citizen board"?

First and foremost, it should be stressed that commissioners are political appointees and not just some sectoral or random slice of the general local public. Their roles vary under the respective mayors who have originally appointed them, and urban observers have already sensed a measurable difference in the way commissions are selected and perceived in the recent transformation from the Tom Bradley era to the Richard Riordan era. Although solid data are underdeveloped, anecdotal evidence[5] gives rise to the conclusion that Mayor Riordan, in his "restructuring" efforts to "turn L.A. around," has been seeking strong conformity from his commissioners to vote his like mind. Independent, solo, nonteam commissioners have already been asked to submit their resignations. There is clearly nothing illegal about this, nor is this beyond the parameters of the charter. What is clear is that political power is being used skillfully to bring forth the "Riordan agenda" through the commission structure and bypassing at times the legislative blessings of the city council and oversight by the electorate. This eventually will alter the balance of power between the mayor and the city council and public decision making emanating out of the commission structure.[6]

Parenthetically, recent controversy has erupted over this struggle, as an increasingly assertive council attempts to hold on to its constitutional charter powers over the city's administrative units and the "stacking" of pro-Riordan commissioners.[7] After 20 years of Bradley's colorful and checkered appointments, the council senses that this is not business as usual. In the past, individual commissioners could be won over by argument, backslapping, and exchange of "chits." Although Bradley removed commissioners through the years for going against his wishes, it was never part of the front-load marching orders that commissioners weighed when vote casting in their boards.

Returning to my line of questioning, then, who are these commissioners, and where do they come from? Furthermore, can one hypothesize about how they may vote on public policy matters, given their individual backgrounds and class status? Mayor Riordan has assembled what I term a *neo-rainbow coalition, based on class, not on race* in his selection of city commissioners. The use of public data supplied by the city clerk's office[8] allows some interesting patterns to emerge. As an aggregate, Mayor Riordan's neo-rainbow commissioner coalition looks like this:

African American	14.693%
Hispanic (Latino)	18.367%
Asian American	8.978%
Female (110)	44.897%
Male (130)	53.061%
Gay and lesbian (5)	2.048%
Total	> 100%[9]

This neo-rainbow coalition, *devoid* of an analysis based on *class,* has a remarkable likeness to the diversity of the city's population. On the basis of the percentage of all *voters* in Los Angeles in the 1993 mayoral election, 18% were African Americans; 8% were Hispanic (Latinos), 4% were Asian Americans, 51% were women, 49% were men, and 5% were gays and lesbians (Simon, 1993). Thus, what can be drawn as preliminary political conclusions regarding the diversity of Mayor Riordan's commission appointments and their reflection of the overall complexities of who Los Angelenos are? What should jump out immediately is that this administration intends to hold together its *electoral* coalition of diverse constituencies, even if that means a solidification of a dominance of Anglo appointments to the city boards and commissions. Anglos constituted 68% of all voters (in 1993) and have been rewarded with 57.9% of all commission appointments. This is ironic, of course, because in the 1990 census, Anglos composed only 35% of the total population of Los Angeles and, on the basis of pure population percentages, are overly represented in this administration's commissions. Furthermore, commission appointments of African Americans and gays and lesbians seem to be a bit below their electoral percentage numbers—perhaps reflecting the declining political strength they will pull in this new mayoral era. Last, Asian Americans and Latinos appear to have done extremely well in commission appointments in relationship to their current electoral composition. Their much higher proportional representation on commissions may indeed allude to new shifting power alliances under this Riordan-governing neo-rainbow coalition.

A more challenging and daunting task is to extract from the data a thesis that the commissioners' class status (and hence "class bias") can be determined by communities of residence. This is based in part on Baudrillard's (1972) topology of the different values of a dwelling (i.e., use value, exchange value, sign value, and symbol value). Baudrillard's scheme is useful in assigning a value to one's dwelling and extrapolating from this measure a thesis about the "value" of the quality on one's neighborhood, the prestige of the area. The use value of a dwelling, and by extension the use value of the surrounding neighborhood, may permit some empirical grounds to hypothesize about an individual's class and social status, absent racial considerations. If use value analysis can lead to a determinate of class status, is there intellectual support to address the notion of class solidarity and social class bias in city commissioners' decision rendering? A large body of classical scholarship has theorized about the support of class solidarity and the existence of a "ruling class" ethos across America (Aptheker, 1960; Domhoff, 1967; Kolko, 1963; Mills, 1956). Clearly, further empirical study is necessary to address the phenomenon of class status and class solidarity in the current roster of Riordan commission appointments and their decisional policy-voting patterns. For the purposes of this chapter, however, by the mapping of the home addresses of each commissioner (minus the Police Commission), some initial assessments may be offered vis-à-vis the expected aggregate voting preference of this group and their expectant preexisting social class bias on the basis of their residential living patterns.

On the basis of an analysis of the 46 existing standing city commissions, 69.5% of all commissioners must statutorily live within the city limits of Los Angeles. Because the charter gives much leeway on membership qualifications for commissioners, and because some commissions were added by ordinance through the years, almost one third of the city commissioners are *not* qualified electors of the city, which means they do not live in the jurisdictional boundaries of the city. It indeed is problematic, although permitted under the charter, to have for example, the majority (three of the five) of members of the Employee Relations Board living as far away as San Diego and Del Mar. It is indeed sad that these commissioners, who must rule on weighty labor matters of the city of Los Angeles, do not have a flavor of the human debate because they are absentee commissioners.

The fault lines of inquiry here reside not just in the logic (or illogic) of having commissioners residing within the city limits, or that by any stretch of the definition that these are citizen boards composed of Angelenos who are vested in this community by paying taxes, breathing the same air, and sending their children to local public schools. Rather, the intellectual inquiry can be expanded to plot (by mapping) the visual depiction of where commissioners actually reside to extrapolate a developing thesis of an inherent class bias among Mayor Riordan's appointments.

A new sophisticated computer mapping program at the University of Southern California's geography lab made it possible to identify by street address the dispersion of city commissioners, first throughout Southern California (see Map 5.1), and second, in the county of Los Angeles (see Map 5.2). Visually striking is the clustering of commissioners' residences evidenced on both these maps. This includes (a) the geographical range of dispersion—not only within Los Angeles County but also in Ventura, Orange, and San Diego Counties; (b) the massing of the cluster toward the west side of the city, the San Fernando Valley, and north of downtown (South Pasadena-Pasadena areas) and a sprinkling within the San Gabriel Valley; and (c) the apparent redlining of city commission appointments from the inner city and South Central and East Los Angeles. Map 5.1 is particularly striking on this point, denoting a literal exclusionary ring around the poorer, inner-city communities from which *no* commissioners have been drawn.

Politically speaking, the end product of each commission's mandate is public policy. The data subset is also surprising in what analyses can be drawn from separately mapping

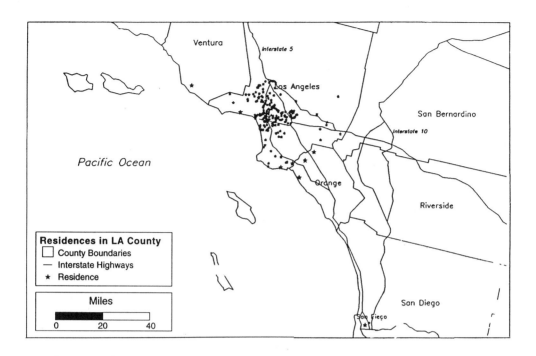

Map 5.1. Los Angeles City Commissioners in Southern California

the chairs and vice chairs of Los Angeles city commissions in Los Angeles County, who control the public policy debate (see Map 5.3). What agenda items are prioritized, the flow and scheduling of debated items, and the parliamentary ground rules for discussion are all left under the charter's housekeeping function with the chairs (in their absence, the vice chairs) of each board. There is, then, an additional element of power invested in these body-elected positions and, therefore, an intellectual need to peel these two positions off for analyzing them independently. Map 5.3 provides a more in-depth mapping of these positions as they fall out according to residential plotting across Los Angeles County. The data appear to suggest an even further concentration and defined *class clustering* of residential patterns, centered in wealthier communities of the city, with an expectation of class bias permeating from within this select subgroup.

Last, it may prove useful to tweak out a grid across Los Angeles County of the household median incomes (based on the 1990 census) and overlay for mapping purposes the residential street address of all the current mayoral city commissioners. Map 5.4 averages out by census tract the aggregate distribution of household median income. Although the individual income for each Riordan-appointed commissioner is unknown, the data suggest in Map 5.4 that by and large these individuals come from relatively affluent income areas and probably bring with them in their public decision-making roles an inherent class bias and prejudice as they determine public policies that affect *all* Angelenos.

72   RETHINKING LOS ANGELES

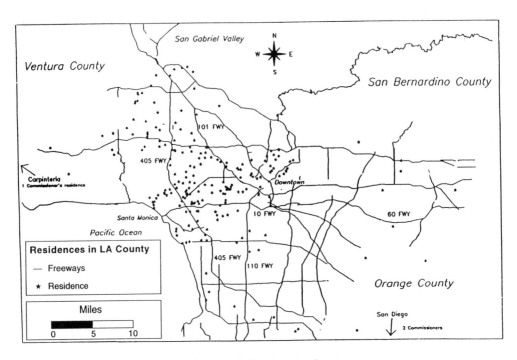

Map 5.2. Los Angeles City Commissioners in Los Angeles County

These four maps, together with the antidotal evidence that Mayor Riordan has assembled a new team with standard marching orders in conformity to the mayor's wishes, unmask yet another antidemocratic formula entombed within the Los Angeles city charter.

In conclusion, we might reflect on how this nightmarish scenario might be improved to salvage the future of urban governability of Los Angeles. If we accept the premise that indeed Los Angeles is governable, how do we wrestle away power to proliferate a new ownership of citizen participation and a new political culture? The stimulus that is required is to develop sufficiently widespread citizen recognition of the current deficiencies and the need to overcome them. This critical mass will not come with piecemeal tinkering at the fringes of power determinants spelled out in the charter. It will come as increased frustration leads to increased participation and other civic virtues that transform a dormant electorate. True charter reform means a radical departure from past deceptions and a reconstructive surgery of elevating average citizens and their neighborhoods to the nucleus of municipal government. Real charter reform is not *the* panacea for all urban ills afflicting Los Angeles; it is, however, the seizure of the attainable, the sine qua non of the possible, and consistent with the legacy of a city that continues to dialectally remake itself as it moves into the 21st century.

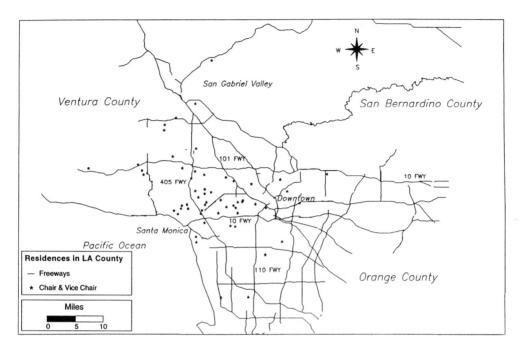

Map 5.3. Chairs and Vice Chairs of Los Angeles City Commissions in Los Angeles County

Notes

1. Following the Watts riots in 1965, several black militant organizations proposed the separation of Watts from the city of Los Angeles. It remained a concept of the far left and never materialized into mainstream politics.

2. This number is the average composition today. Some commissions have as few as 3 members; others, such as the Productivity Commission, have 15.

3. Since 1938, it has been customary for commissioners to offer their resignations when a new mayor takes office to permit a new administration to "implement their campaign promises and agenda."

4. Early administrative boards were widely used between 1900 and 1920 to check bureaucratic and administrative power, out of distrust for "urban professionals" who lost contact with the average citizen.

5. This is based on discussions with several staff members at the chief legislative analyst office and staff members of the city council.

6. Actually, a few years ago, by a literal slip of the pen, Mayor Bradley set into motion what was to become Proposition 5, which passed overwhelmingly. It permits the city council to overrule any decision made by any city commission, usually within 5 working days and by a supermajority of council votes. Although rarely used yet, it may become a tool the council will use to weaken the mayor's policy desires.

7. Such stacking is hardly to be confused with the stacking of the Supreme Court by President Franklin Roosevelt with sympathetic New Deal jurists!

8. A note on data retrieval and analysis is necessary here. The city clerk's office has on public record all commissioners listed by appointed boards, designation of chair and vice chair, home addresses, and zip codes. They refused to provide the supplemental personal information on the five-member Board of Police Commissioners, claiming state law, for safety reasons, prohibits release of this information. Subsequent data analysis and mapping for this chapter should thereby reflect this variable. Furthermore, as of October 1994, Mayor Riordan had made

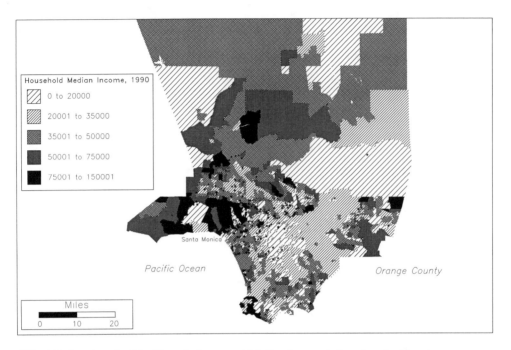

Map 5.4. Los Angeles City Commissioners and Median Income in Los Angeles County
SOURCE: Data on household median income are from U.S. Bureau of the Census (1990).

245 appointments of a total of 272 city commissioners. Existing vacancies were therefore not computed into the analysis.

9. Obviously, the data reflect categories that have been double counted on the basis of an individual's gender, ethnicity, and sexual orientation, producing a universe greater than 100%.

References

Allswang, J. M. (1991). *California initiatives and referendums, 1912-1990.* Los Angeles: California State University, Edmund G. "Pat" Brown Institute of Public Affairs.

Aptheker, H. (1960). *The world of C. Wright Mills.* New York: Marzani & Munsell.

Banfield, E. C., & Wilson, J. Q. (1963). *City politics.* Cambridge: Harvard University Press and MIT Press.

Baudrillard, J. (1972). *Pour une critique de l'economie politique du signe* [For a critique of the symptoms of the political economy]. Paris: Gallimard.

Bollens, J. C. (1963, December). A history of charter reform efforts. In J. C. Bollens (Ed.), *A study of the Los Angeles City Charter.* Los Angeles: Town Hall.

Charter of the City of Los Angeles (Rev. ed.). (1973). Los Angeles: Building News. (Adopted 1925, January 22)

Citizen Ad Hoc Working Group for Charter Reform. (1993, January). *City government for the 21st century.* Draft document.

Connery, R. H., & Caraley, D. (Eds.). (1969). *Governing the city.* New York: Praeger.

Costikyan, E. N., & Lehman, M. (Eds.). (1972). *Restructuring the government of New York City: Report of the Scott Commission Task Force on Jurisdiction and Structure.* New York: Praeger.

Deverell, W., & Sitton, T. (Eds.). (1994). *California progressivism revisited.* Berkeley: University of California Press.

Domhoff, W. G. (1967). *Who rules America?* Englewood Cliffs, NJ: Prentice Hall.

Hays, S. P. (1992). The politics of reform in municipal government in the progressive era. In D. Judd & P. Kantor (Eds.), *Enduring tensions in urban politics.* New York: Macmillan.

Judd, D. R., & Swanstrom, T. (1994). *City politics: Private power and public policy.* New York: HarperCollins.

Key, V. O., & Crouch, W. (1939). *The initiative and referendum in California.* Berkeley: University of California Press.

Kolko, G. (1963). *The triumph of American conservatism.* New York: Free Press.

Lowi, T. (1964). *At the pleasure of the mayor.* New York: Free Press.

Mills, W. C. (1956). *The power elite.* New York: Oxford University Press.

Mowry, G. (1951). *The California progressives.* Chicago: Quadrangle.

National Advisory Commission on Civil Disorders (Kerner Commission). (1968). *Report of the National Advisory Commission on Civil Disorders.* Washington, DC: Government Printing Office.

Report of the Los Angeles City Charter Commission: City government for the future. (1969, July). Los Angeles: Los Angeles City Charter Commission.

Ries, J. C., & Kirlin, J. J. (1971). Government in the Los Angeles area: The issue of centralization and decentralization. In W. Z. Hirsch (Ed.), *Los Angeles: Viability and prospects for metropolitan leadership.* New York: Praeger.

Schockman, H. E. (1990, February 4). Nonpartisan city elections produce a leaderless ship. *Los Angeles Times,* p. M5.

Schockman, H. E. (1991, August 4). The city charter that doesn't work. *Los Angeles Daily News,* p. VPT1.

Schockman, H. E., & Majd-Sadjadi, Z. (1994). After NAFTA. *California policy choices, Vol. 9.* Sacramento: University of Southern California, School of Public Administration.

Simon, R. (1993, June 9). Exit polls find voters sought lesser of two evils. *Los Angeles Times,* p. A1.

Stone, C. N. (1989). *Regime politics: Governing Atlanta.* Lawrence: University of Kansas Press.

U.S. Bureau of the Census. (1990). *County and city data book: 1990.* Washington, DC: Government Printing Office.

Community Policing

MARK KROEKER
Deputy Chief, Los Angeles Police Department

Police officers are committed people doing an extraordinary job. Men and women join the Los Angeles Police Department desiring to serve their city and "make a difference." In Operations-South Bureau, our aim is to make a difference by focusing on three fronts: (a) reduce crime and the *fear* of crime, (b) improve the quality of life in our neighborhoods, and (c) improve our C.P.R.

We are focusing our resources to combat the violent and gang-related crime that victimizes whole communities. The neighborhood's quality of life is raised by reducing crime and removing the blight from crime, such as graffiti. C.P.R. translates to *community police relations* and, like its medical use, we must administer a fresh breath to the community we represent. Through positive and constructive relationships with community, we can produce remarkable results. To administer C.P.R., we are actively integrating true "community policing" through our South City divisions. This is the first step toward "community government." No longer can the "us-them" relationship be tolerated. Our officers are continuously working with members of their communities to identify and prioritize crime problems and develop sustainable solutions. These strategies contribute to a better Los Angeles and operationalize our motto, "to protect and serve."

Los Angeles and the Future 6
Uprisings, Identity, and New Institutions

HARLAN HAHN

Los Angeles: The Contextual Framework

The history of Southern California has been permeated by a continuing debate about the nature and meaning of human differences. For many years, racial and ethnic minorities were the principal objects of public attention; hence, most statements on this issue have reflected an implicit racist discourse. Ironically, minorities constituted a majority among the original residents of the village that would eventually become Los Angeles; most of the founding members of the settlement in 1781 were persons of color (Bond, 1936, pp. 3-6). Gradually, however, the texture of the community changed as whites became the dominant portion of the population. Perhaps the clearest public evidence of racist attitudes was revealed by speeches delivered at the California constitutional convention in 1849. The delegates expediently defeated a proposal to prohibit the immigration of African Americans and adopted a ban on slavery in the state. But W. M. Shannon, the author of the latter resolution, apparently achieved success in these debates with his argument that the promotion of "domestic life" required the labor of everyone, including "baboons, or any other class of creatures" (quoted in Almaguer, 1994, p. 37).

AUTHOR'S NOTE: Appreciation is expressed to Judson Lance Jeffries, H. Eric Schockman, and Jose Gomez for their assistance in the research and for their comments on earlier drafts of the manuscript.

During most of its subsequent history, Los Angeles could be described as a product of centrifugal tendencies (Fogelson, 1967). In general, the dynamics of local development have emerged less from a defining nucleus than from an outward thrust at its margins. Part of the sprawling configurations of Los Angeles probably can be traced to the tenets of progressivism, which won a titanic battle with the Southern Pacific Railroad at a crucial point in the transition of the municipality from a sleepy pueblo to a bustling commercial center (Mowry, 1951). These reformers sought to decentralize governmental institutions as a means of preventing the reemergence of political bosses. Thus, the city remained firmly within the grip of a "Yankee-Protestant" political ethos that "expressed a common feeling that government should be in a good part an effort to moralize the lives of individuals," whereas the economic system would provide an outlet for self-aggrandizing instincts (Hofstadter, 1955, p. 9). Beneath the geographic dispersion of public authority lurked a businesslike faith in the principle of professional specialization, or a division of work that is even displayed in the titles of suburbs, which range from prosaic labels such as Industry and Commerce to names that connote idyllic residential retreats such as Pacific Palisades and Woodland Hills.

Los Angeles also embraced nonpartisan at-large elections that have traditionally placed minority groups at a disadvantage. These planks in the progressive platform were adopted at a time when racial and ethnic minorities were confined to segregated neighborhoods and a virulent anti-Semitism was spread in the pages of the *Los Angeles Times* as it crusaded against both the reformers and socialist candidates. City voters in 1925 endorsed a return to councilmanic districts. In 1926, however, a referendum annexing the community of Watts was approved with the support of the Ku Klux Klan, "thereby preventing a Black-dominated local government" (Sonenshein, 1993, p. 27). In addition, faint traces of a fear of miscegenation were revealed by the decision of the Santa Monica city council in 1922 to close an African American dance hall and to prohibit the construction of another amusement center and a bathhouse (DeGraaf, 1970, p. 348). African Americans continued to be limited to a crowded section of Los Angeles by the combined force of white intimidation and restrictive covenants, which were not declared unconstitutional until 1948. But these residents did not compose the only minority that was subjugated by white prejudice. Latinos "had the worst housing of any group" (DeGraaf, 1970, p. 339). Native Americans were forced to live on barren reservations. And dominant animosity toward Asians was dramatically revealed by policies ranging from the Chinese exclusion laws to the incarceration of Japanese Americans during World War II.

The influences unleashed by the war marked another watershed in the history of Los Angeles. As jobs in defense industries became a primary route to upward mobility, the metropolis confronted a shortage of young, white, nondisabled, heterosexual men, who were serving abroad in the military. Their absence allowed many women, gays, persons with disabilities, older workers, and minorities to enter the labor force for the first time. Thousands of African Americans and Latinos migrated to the city to find employment, but their search for adequate housing was thwarted by the renewed hostility of white citizens (Davis, 1992, p. 163). After the war, they faced the bleak prospect of unemployment and the lack of a suitable place to live.

By contrast, white Angelenos became major beneficiaries of the suburbanization that flourished in the wake of World War II. Many perceived the burgeoning minority population as a threat to cherished values, and they responded by fleeing in droves to the outskirts of the municipality. The trend, of course, accelerated the decentralization of governmental authority in the region.

Perhaps most significant, these developments were artificially supported by public policies that favored suburbanites at the expense of the city and the disadvantaged residents they left behind. "White flight" to the suburbs was almost literally paved by federal subsidies. Government programs, for example, provided low-interest loans to veterans and civilians to enable purchasers to fulfill the American dream of owning a home. Defense contracts were awarded by public officials to local corporations without a careful investigation of employment practices that discriminated against women and minorities. And a massive highway program was approved by Congress to build convenient freeways that permitted commuters to move each day between comfortable bedrooms and urban work sites. Travel in a sealed glass-and-metal vehicular cocoon apparently screened their view of the desperation and decay surrounding them in the central city. Furthermore, officials in Los Angeles devised a scheme to assist communities that wanted to incorporate without paying increased taxes to meet their needs. Under the "Lakewood plan," suburbs contracted with the county to obtain local services, thereby shifting the financial burden to taxpayers throughout the metropolitan area (Davis, 1992, pp. 165-166). As a result, white homeowners were allowed to seize control of their own territory within cohesive suburbs at relatively few personal costs.

Gradually, emerging trends in postwar Los Angeles seemed to signify a subtle shift in ideological values. Freedom was no longer determined solely by the availability of local opportunities for personal enhancement or fulfillment through mutual cooperation. Instead, liberty was equated with autonomy that defined the relentless pursuit of individual self-interest, unhindered by social or governmental constraints, as the epitome of this conception. As the community moved from the turbulent 1960s to the conservative milieu of the 1970s and 1980s, the prevailing ethos of Los Angeles began to display extreme forms of polarization, as well as individualism and insularity. The implicit nonverbal posture of residents meeting for the first time on urban streets appeared to change from a solicitous concern about their mutual well-being to a wary suspicion of the other person's motives. The predominate means of communication between strangers in Los Angeles seems to consist of reading each other's bumper stickers and license plates on the freeways.

The Civic Uprisings: Minority Goals

The vast chasm that separates the white majority and racial and ethnic minorities in Los Angeles was dramatically revealed by the civic uprisings of 1965 and 1992. Both these events seemed to represent an emancipatory struggle for social and political liberation in an effort to move beyond pervasive alienation in minority neighborhoods to a variant of the autonomy that permeates the white community. In many respects, the uprisings also resembled the type of "collective bargaining by violence" that occurred in many European cities before the invention of democratic institutions (Hobsbawm, 1965). Lacking any formal method of communicating their grievances to urban elites through elections or representation, throngs of ordinary citizens occasionally fomented disorders in the streets so that a delegation from the unruly masses would be summoned to meet with local officials. After their negotiations, city authorities often would announce a compromise or settlement to the crowds when the militia was called to quell the disturbance. Racial and ethnic minorities in Los Angeles also felt disenfranchised and unrepresented as a result of their inability to achieve significant goals through traditional strategies such as rational arguments, voting, participation in political parties, campaigns for favored candidates, lawsuits, lobbying, and mass demonstrations. But the important point is that the uprisings of 1965 and 1992 did not occur until after such tactics had failed to achieve major political objectives. The view of these incidents as manifestations of a "politics of last resort," therefore, perceives them as the cumulative result of arduous years of struggle rather than as simple or sudden outbursts of protest.

In general, the differences between the Los Angeles uprisings of 1965 and 1992 seemed to be overshadowed by their similarities. In both years, the precipitating incidents that sparked the violence signified long-standing and deep-seated grievances about the imposition of police authority in Los Angeles. Second, Latinos were more extensively involved in 1992 than in 1965, but this trend only reflected both the changing composition of racial and ethnic enclaves and the basic resentment against discrimination that is shared by minority groups. Third, much of the violence in 1992 was directed at Korean American stores, but this pattern also indicated demographic changes in business ownership; more important, it obscured the fundamental fact that the attacks on retail establishments in both years were selective rather than random or senseless. The principal goal underlying both actions was encompassed by a massive demand for the restructuring of social and political institutions so that ghetto residents could control their own fate.

The occurrences in Los Angeles of 1965 and 1992 were not unprecedented. White residents had frequently used violence as a means of compelling African Americans, Latinos, Chinese Americans, and Japanese Americans to occupy a subordinate position in the civic order. Perhaps the major initial feature that differentiated subsequent events from these earlier conflicts was the pivotal role of the police in the circumstances that precipitated the uprisings in South Central Los Angeles as well as similar developments in East Los Angeles following the death of journalist Ruben Salazar by police action at the National Chicano Moratorium in 1970. To average citizens, police patrols probably represent the most tangible and direct means by which governmental authority is imposed on their lives. In minority enclaves, this influence has been undermined by a legacy of hostile and abrasive encounters with the police that stemmed in part from official law enforcement policies (Feagin & Hahn, 1970). Thus, the public display of opposition to police practices signaled that officials outside the community were no longer in control of local activities. In a fundamental sense, the uprisings signified the deliberate defiance of established sources of municipal authority and the continuing repudiation of attempts to reimpose that authority on disadvantaged neighborhoods. Hence, the meaning of these actions also reflected a purposive challenge to city authorities that extended beyond the mere expression of discontent.

The local response to police behavior at the precipitating incidents marked another crucial transition to what has been called *street governance* (Hahn & Holland, 1976). Within a few hours, the social order had been turned upside down. For a short time, in the absence of police control, political sovereignty was brought down from the remote corridors of officialdom to throngs of ordinary people on the streets. These crowds constituted the basic rule-making body in South Central Los Angeles at least until conventional forms of governmental authority were reimposed. They represented a shifting assembly of diverse citizens who acted on the basis of an unspoken consensus. The essential thrust of their conduct was participatory. For the first time, persons who were active in the civic uprisings had an opportunity to become involved in a decision-making process that might enable them to shape their own destiny.

Many of the rules formed on the streets during the uprisings were not confined to a mere renunciation of prior laws and regulations. In 1965, "Soul Brother" signs designating businesses owned by African Americans often were respected—when they were authentic. In both years, innocent bystanders were protected by members of the crowds. During the period of street governance, ghetto residents were engaged in the complex and difficult process of arriving at a new understanding of principles that they believed ought to guide political decisions. In essence, they were seeking the same right to self-determination that had been granted to insular communities elsewhere.

Perhaps the most significant evidence about the normative standards that emerged in the uprisings was revealed by the relatively selective nature of assaults on retail establishments. This pattern was shaped by attitudes that had formed during a long period. Undoubtedly, some of the looting was prompted by opportunity. Some of the arson may have been motivated by aimless rage. But informants who had lived in these neighborhoods for many years reported that whereas most of the businesses that were spared from such attacks had previously enjoyed a good relationship with local residents, a major share of the buildings that had been burned or looted were owned by persons with unfavorable reputations (Hahn, 1992). The actions of participants in the uprisings revealed identifiable efforts to distinguish between businesses that had maintained positive relations with the community and those that had engaged in abrasive or exploitive practices.

Most outside observers failed to recognize these distinctions, and they focused instead on the illegality of arson and looting. From a perspective that regards the uprisings as the product of long-standing grievances and as a manifestation of street governance, however, this conduct can be interpreted as signifying a redefinition of social and political values. During these events, personal needs were granted a higher priority than the preservation of private property. Consumer goods that had long been coveted were confiscated by methods that implied the endorsement, if not the explicit authorization, of crowds in the neighborhood. In many respects, the looting resembled a form of expropriation that challenged the legal foundations of the prior consensus concerning the distribution of property (Quarantelli & Dynes, 1968). More important, these actions revealed a fundamental belief that major social and economic problems cannot be remedied without a significant redistribution of public and private resources. Expropriation, whether it is carried out by duly constituted government leaders or by crowds that have temporarily seized control of an area, is basically a form of redistribution; although this strategy has usually been omitted from debates about public policy, redistributive principles have long been a crucial part of the agenda of deprived and disadvantaged groups seeking massive social changes.

Perhaps the major contrast between the Los Angeles uprisings of 1965 and 1992 was revealed by actions taken in the wake of each event. In 1965, public concern about the violence in Watts enabled Los Angeles to overcome the effects of chronic decentralization and to jump momentarily from last to first in a ranking of the nation's four largest cities by the amount of antipoverty funds that they received from the federal government (Greenstone & Peterson, 1968, pp. 287-288). These appropriations were not intrinsically redistributive because they did not encompass efforts to alter tax laws so that resources would flow from wealthy segments of society to the urban poor. But they did indicate the strong belief of local and federal officials in the ability of government to solve persistent social problems by creating innovative programs to assist disadvantaged groups.

In 1992, by contrast, the proposals that were offered in response to the uprisings centered almost exclusively on attempts by former baseball commissioner Peter Ueberroth to persuade major corporations to contribute voluntarily to his committee to "Rebuild Los Angeles." This focus seemed to denote a dramatic shift from a faith in government to a reliance on the private sector that was consistent with the growing adherence to the concept of autonomy, or a resistance to the imposition of governmental restraints on individual conduct. Without a strong political ally, however, private initiatives appeared to be a relatively ineffectual means of alleviating the conditions that had ignited the uprisings. As a result, Los Angeles confronted an uncertain fate devoid of any tangible plan to avoid future uprisings except the vague hope that—somehow—the events of 1965 and 1992 would not be repeated.

The Evolution of the Present Agenda

The declining confidence in the public sector that occurred between the 1960s and the 1990s also appeared to affect the perspective of social scientists. Significantly, the 1992 uprising failed to stimulate renewed consideration of the concept of community control, or the delegation of decision-making authority to urban neighborhoods, that had been proposed by researchers in the aftermath of the violence in 1965 (see, e.g., Altschuler, 1970; Frederickson, 1973; Kotler, 1969). In many respects, this plan seemed to be a logical response to the demand for participation and self-determination that emerged during both events.

The aims of persons who were active in the uprisings also may have been shaped by the model of local government that was evident on the edges of Los Angeles. In part, they simply wanted what white suburbanites already had. The objectives of both groups reflected the desire for a restructuring of political institutions to facilitate increased participation in formulating policies that would promote distinctive values and lifestyles within relatively small and homogeneous jurisdictions. In the harmonious atmosphere of suburban politics, decisions are usually based on a local consensus, and the opportunity to participate is readily available to almost anyone who wishes to become involved. Ironically, however, analysts did not acknowledge the similarity of the goals of ghetto residents in the 1992 uprising. The failure to grasp the analogy between the aspirations of suburbanites and those of inner-city residents seemed to denote a corresponding reluctance to recognize the benefits conferred on whites and the disadvantages imposed on minorities by existing governmental institutions.

In several respects, community control seems to be an attractive option in the creation of a new system of intra-urban governance (Hahn, 1994). The formation of neighborhood governments could redress the inequities that exist between suburban whites and urban minorities; it would permit ghetto residents to shape their own destiny. Furthermore, many of the city's most pressing problems, such as air pollution, transportation, and crime, could be ameliorated by fostering an enhanced sense of community. Increased use of carpools and mass transit would reduce smog as well as traffic congestion on the freeways, and expanded neighborhood surveillance and self-policing might have a greater impact on crime than law enforcement officers. Perhaps innovative institutions can be formed to facilitate a recognition that the fate of Los Angeles depends on the promotion of cooperative endeavors instead of the unrestricted pursuit of individual autonomy.

The principle of community control, however, also raises a classic dilemma concerning the design of governmental organizations. On the one hand, by delegating decisions to relatively small homogeneous areas in which citizens can gain access to the political process, decentralization generally promotes increased participation. On the other hand, to mobilize the resources necessary to secure significant accomplishments, centralization is usually desirable. If plans for community control included a major transfer of authority and an appropriate redistribution of government funds, they may be successful in combating the growth of alienation in minority neighborhoods. But politicians ordinarily are reluctant to undermine their own influence by granting substantial power or financial support to subordinate institutions. Efforts to solve the vast social problems in Los Angeles also require major public expenditures that can be provided only by a centralized decision-making structure. Decentralization may tend to promote expanded political activity, but in poverty-stricken ghettos and barrios, increased participation often appears to be a luxury that residents cannot afford.

Some observers, of course, might argue that the formation of organizations representing disadvantaged groups is unnecessary in Los Angeles. As an African American candidate for mayor who defeated popular white opponents in several elections, Tom Bradley repeatedly gained a larger percentage of the white vote than did African American politicians elsewhere (Hahn, Klingman, & Pachon, 1976). Jackson, Gerber, and Cain (1994, p. 283) also found that African American residents of Los Angeles felt closer to whites and Latinos than did a national sample of African Americans. As a result, Sonenshein (1993) has contended that the success of Bradley represented the prototype of a black-white alliance, especially a black-liberal Jewish coalition, that could be replicated in subsequent years.

Several caveats, however, must be attached to the enthusiasm generated by this scenario. First, particularly during the later years of Bradley's incumbency, emphasis increasingly was focused on programs that provided major advantages for business interests instead of for poor sectors of the population (Sonenshein, 1993, pp. 163-175). Most of Bradley's proposals concentrated on indivisible public policies that affected the city as a whole rather than on distributive or redistributive projects offering direct benefits to disadvantaged groups (Hahn, 1976). This orientation, of course, did not diminish Bradley's support among African American voters, but it indicates the possibility that his election merely reflected a vicarious triumph comparable with the psychic satisfaction that earlier ethnic groups had derived from elevating one of their members to high public office without receiving tangible economic rewards for their efforts. Similarly, the African American-liberal white alliance that backed Bradley may have been more important in his precampaign drive to gather political contributions than in his quest for electoral support. In 1969, for example, the percentage of the white vote that he received at the polls probably represented a smaller share of his margin of victory than the proportion of the $1 million that white benefactors donated to his campaign fund (Sonenshein, 1993, pp. 90-94). Latino politicians also have appeared to gain success in Los Angeles primarily on the basis of the financial contributions that they raised through their contacts in the state legislature in Sacramento (Skerry, 1993).

In addition, surveys during the 1969 election revealed that opposition to Bradley was related to the anxiety that whites were falling behind African Americans in striving to achieve major social values (Pettigrew, 1971; Sears & Kinder, 1971). Additional data from a 1973 survey confirmed that white support for Bradley's opponents in Los Angeles was based less on a perceived threat to personal interests than on a "symbolic racism" that combined antiblack sentiments with the "traditional moral values embodied in the Protestant Ethic" (Kinder & Sears, 1981, p. 416). Other survey evidence on fair housing referenda in California and elsewhere disclosed that among whites who took a liberal position on other racial issues, the effects of sophistication and social desirability may have influenced upper-class voters to record greater support for these measures than their counterparts at lower levels of socioeconomic status (Hahn, 1968). White attitudes, therefore, have revealed increasingly subtle and covert forms of discrimination instead of blatant expressions of racial and ethnic prejudice.

Finally, the victory of white Republican Richard Riordan in his 1993 mayoralty campaign against Michael Woo, the first Asian American member of the city council, seemed to signal the collapse of the coalition that had supported Bradley. Exit polls revealed that although Woo received 86% of the votes of African Americans and 61% of the ballots cast by Asian Americans, Riordan was endorsed by 85% of the white voters; the vote among Latinos, Jews, and women was split almost evenly between the two candidates (Preston, 1993, pp. 34-35). Voting in Los Angeles has continued to reflect racial polarization, rather than a shifting reconfiguration of preferences between African American and white segments of the electorate. In the 1993 election, white support for a biracial coalition virtually disappeared, but the prospect of forging an alliance among disadvantaged groups in Los Angeles seemed to remain a viable, though problematic, alternative. A study of African American-Latino antagonism found that 78% of the Latinos felt that African Americans did not have "too much political power" in the city and that 87% of the Latinos and 81% of the African Americans disagreed with the view that local relations between the two groups were "getting worse" (Oliver & Johnson, 1984, p. 77). Hence, most African Americans and Latinos seemed to be receptive to the possibility of improving political relationships between their respective communities in Los Angeles. Moreover, the potential for a multiethnic alliance was not adversely affected by the civic uprising. A 1992 survey revealed that although Korean American businesses were major targets of the violence in Los Angeles, both Asian Americans and Latinos were evenly divided by the question of whether the uprising represented a form of protest or a means of "engaging in looting and street crime" (Bobo, Zubrinsky, Johnson, & Oliver, 1994, p. 111).

The formation of a successful coalition among disadvantaged groups in Los Angeles, however, probably cannot be accomplished without a major restructuring of electoral institutions. There is little doubt that "Los Angeles' system of coupling district elections with nonpartisanship has contributed to a decentralized politics" (Carney, 1964, p. 112). Attention, therefore, might appropriately focus on additional reforms that would combine the advantages of a centralized mechanism for mobilizing needed resources with plentiful opportunities for citizens to become involved in the decision-making process.

Some of these changes have been marked by plans launched after the passage of the Voting Rights Act of 1965. These programs have focused on efforts to form political jurisdictions and legislative districts that would ensure the representation of racial and ethnic groups as well as on actions to increase voter registration and turnout in minority communities. In Los Angeles, attempts by disadvantaged groups to gain self-determination in separate suburbs have produced mixed results. Although an early attempt by Latinos to incorporate portions of East Los Angeles was defeated (Skerry, 1993, p. 78), gays and lesbians joined with aging tenants in 1984 to create the city of West Hollywood (Moos, 1989). Between 1985 and 1987, however, Latino representation on the Los Angeles City Council increased from zero to two in part due to a lawsuit under the Voting Rights Act that stimulated reapportionment to consolidate Latino strength in particular districts. In addition, research on the 100 top elected positions in Los Angeles County between 1960 and 1986 disclosed that although Jewish representation has increased steadily, initial gains by African Americans and Latinos through reapportionment and vacancies subsequently reached a plateau (Guerra, 1987). Representation of racial and ethnic minorities by geographical areas, therefore, seemed to impose an invisible ceiling or quota on disadvantaged groups that prevented them from acquiring an influence in local government proportionate to their share of the population.

To avoid the dilution of minority preferences in elections and legislative bodies, Guinier (1994) has advocated cumulative voting as part of a third generation of proposals to implement the principles of the Voting Rights Act. Under this plan, each voter might be given a number of ballots equivalent to the number of legislative seats contested in an election; the voter would be free to distribute votes between the candidates in any manner that the voter chooses. As a result, disadvantaged groups would be able to register the intensity of their support for preferred office seekers as a means of enhancing their influence in legislative chambers. This proposal represents an important step toward solving the inherent dilemma produced by the presence of electoral polarization and persistent minorities within a political system based on majority rule. It also constitutes a nonterritorial alternative that would ameliorate some of the moral contradictions involved in basing the representation of minority groups on geographic districts that were shaped by an egregious history of residential segregation. But cumulative voting does not resolve the additional problems created by the disproportionate power of money in political campaigns and by the need for authentic representation that would faithfully serve the needs of minority groups in the policy-making process. Presumably, for example, minority legislators elected by cumulative voting could be persuaded by massive contributions from white donors to favor the interests of dominant groups instead of their own constituents, and voters might similarly be prevented from recognizing the defection of such leaders by the influence of campaign propaganda. As a result, there seems to be a pressing need to devise a remedy for the barriers facing disadvantaged groups founded on a new conceptual perspective.

Perhaps to a greater extent than social scientists commonly realize, research paradigms may have an important effect not only on how people think about an issue but also on the solutions that they propose. A clear example of this phenomenon is evident in the vast literature on African Americans, Asian Americans, Native Americans, Latinos, women, gays and lesbians, older persons, and people with disabilities. Most analysts do not approach the study of these groups within a comparative framework; instead, they treat them as separate and independent entities. Correspondingly, such studies tend to stress the unique and specific concerns of each of these portions of the population instead of the common objectives that they share. By instilling a mode of thought that emphasizes the discrete characteristics of disadvantaged segments of society, perhaps researchers have unwittingly impeded the process of coalition formation that might yield a general remedy for their problems. The following discussion, therefore, introduces an alternative model focusing on the mutual goals of groups that have been disadvantaged by race or ethnicity, gender, sexual orientation, age, and disability. Although this outline is admittedly sketchy, it is designed to foster the task of rethinking the challenges that confront major cities such as Los Angeles.

The Redefinition of Identity and Issues

One of the most important goals of social movements in the second half of the 20th century is signified by the effort to translate previously devalued attributes such as skin color, gender, age, and disability into a positive sense of personal and political identity. In general, these strivings encompass both an internal struggle to achieve increased self-esteem and public action to advance the collective aspirations of the group. Anspach (1979) has traced the emergence of the disability rights movements, for example, to the growth of political activism among persons who had formed a favorable self-concept and a critical view of society. The constant interplay between the psychological and social dimensions of these movements shapes their attempts both to attain major political objectives and to facilitate the transition from individual feelings of inferiority and shame to an affirmation of dignity and pride. People who develop high self-esteem are especially apt to become active in promoting the aims of disadvantaged groups, and participation in such endeavors tends to bolster their self-image. As a result, the effects of social movements on their members may be almost as important as the impact that they exert on outside observers.

The continuous interaction between the personal and social facets of movement politics is usually revealed by a series of developments. Initially, individuals in a disadvantaged segment of society must be persuaded to abandon a strategy of "passing," by which many of them have previously sought to disguise their affiliation with the group to assimilate with the dominant majority. Among gays and lesbians, this phase frequently is described as "coming out." For persons with perceptible differences such as skin color or gender as well as the visible signs of aging and disability, it commonly entails a redefinition of physical traits. During this process, which is also marked by the repudiation of prevalent stereotypes, members of these groups frequently seek to attach a positive meaning to attributes that were once a source of oppression; they adopt this new interpretation of their salient characteristics as a primary basis for political identity. Perhaps the clearest example of this process, which is crucial to the mobilization of a social movement, is the emergence of the concept that "black is beautiful" in the early 1960s. Similar developments, of course, were revealed by the growth of a militant feminism and by an increasing sense of pride among gays and lesbians, aging individuals, and people with disabilities. The social taint that had previously constrained their political activism usually becomes the major target of their common endeavors; stigma is often reconceptualized as a form of prejudice or as a manifestation of racism, sexism, homophobia, ageism, and physicalism. As a result, participants in the struggle to combat these forms of discrimination learn that the origins of the principal barriers they confront are located in the external environment of public attitudes rather than within themselves.

The task of redefining individual characteristics often is supported by attempts to reshape public discourse about the nature and meaning of human differences. Through a process that is described in the feminist movement as "consciousness raising," women begin to identify gender rather than other personal attributes as the primary source of the problems they face; again, comparable interpretations are promoted by rhetoric at meetings or rallies of every group that emphasizes the goals that unite the movement. Perhaps the principal message conveyed by subsequent discussions, however, reflect that all disadvantaged groups are fundamentally defined by social and cultural influences rather than by biological properties. As a result of extended dialogue about these matters, these groups may eventually form distinct constituencies, or blocs of voters who select candidates according to their positions on issues relevant to their segment of the electorate. Moreover, the formation of such constituencies, the discrimination to which these groups have been subjected, and the constancy of their characteristics constitute a strong justification for providing political representation on the basis of race or ethnicity, gender, sexual orientation, age, and disability.

In a fundamental sense, the social movements that have emerged in the second half of the 20th century signify an effort to translate the lived, embodied experience of disadvantaged groups into theory and practice. Most significant, this goal has seemed to denote a concomitant shift from the pursuit of liberty or even self-determination to a quest for equality and justice.

The struggle for equality has focused thus far primarily on attempts to end discrimination. These efforts have resulted in the passage of major national legislation such as the Civil Rights Act of 1964, prohibiting discrimination on the basis of race or ethnicity and sex; the amendments to the Age Discrimination in Employment Act of 1967, forbidding mandatory retirement; as well as Section 504 of the Rehabilitation Act of 1973 and the Americans With Disabilities Act of 1990, outlawing discrimination against persons who are disabled. No federal antidiscrimination law has been enacted as yet to protect the rights of gays and lesbians. Moreover, Minow (1990) has argued that the interpretation of civil rights statutes should reflect a "situated perspective" encompassing the viewpoint of the observed as well as the observer. Feminist scholars also have pointed out that supposedly neutral legal rules to curb gender discrimination often bear the implicit traces of patriarchal values, and they have urged that interpretations of the concept of equality ought to be founded on an explicit recognition of the disadvantaged status of women (Rhode, 1989). A similar criterion could be applied to policies involving minorities demarcated by race, ethnicity, sexual orientation, age, and disability. Hence, the effect of public or private actions on the advantages or disadvantages bestowed on these groups can be adopted as a standard for assessing the extent to which they have gained equal rights.

In addition, however, the goals of disadvantaged groups may provide a basis for evaluating the concept of justice (Young, 1990). An enhanced appreciation of their objectives cannot, of course, be expected to yield a complete resolution of the tension between altruism and self-interest. Yet much can be gained by an increased recognition that in addition to the understandable pursuit of economic ambitions in a capitalist system, these groups are united by an overarching struggle against discrimination. The effort to promote an expanded awareness of the common effects of disadvantage also could eventually lead to the formation of coalitions based on similar responses to major problems in life such as forming or maintaining personal relationships, raising children, securing satisfactory work, and coping with vulnerability to chronic disability and increased longevity. A crucial prerequisite to the development of a remedy for the obstacles confronting social movements, therefore, appears to require a reconceptualization that focuses increased emphasis on the parallel experiences of these groups. Moreover, the creation of institutions and programs to foster the coalescence of disadvantaged segments of society may avoid many misconceptions resulting from the relative lack of class consciousness in American society. Although economic interests might ultimately emerge as a factor of transcendent importance in political controversies, the most appropriate foundation for the unification of disadvantaged segments of the electorate seems to revolve about their quest to implement crucial principles of equality and fairness. Perhaps even more significant, by forging effective alliances, the combined strength of groups that are the targets of discrimination based on race, ethnicity, gender, sexual orientation, age, and disability would constitute an overwhelming majority of the electorate.

The aspirations of disadvantaged groups also can be promoted by a redefinition of political issues. This analysis is based on the assumption that political concerns often emerge naturally from extensive social interactions within the community. Also incorporated within this approach is the feminist principle that "the personal is political." In many respects, this axiom implies that attempts to dismantle the rigid division between private and public spheres of activity might uncover important new social problems that have been excluded from the civic agenda by the influence of governing elites. Some evidence concerning this possibility is provided by surveys of visiting patterns that have been conducted in Los Angeles since the end of World War II.

Contrary to the image of an impersonal city, these data disclosed that most white residents of Los Angeles visited with family members at least once a week (Greer, 1972). In general, contacts with neighbors and coworkers have reflected strong patterns of racial and ethnic polarization, but other results indicate that kinship visiting is more prevalent among African Americans, and especially Latinos, than among white Angelenos (Bengtson, Cutler, Managen, & Marshall, 1985; Grebler, Moore, & Guzman, 1970). Among middle-aged or older adults, whites are more apt to visit with friends and neighbors, whereas African Americans and Latinos are more likely to talk with relatives (Dowd & Bengtson, 1978). Although little research has been conducted on the content of these conversations, the findings appear to support the speculation that family problems or concerns about the development of children may form the basis for frequent visits in Los Angeles. Hence, providing a public forum for the discussion of such issues could promote increased political involvement among diverse segments of the population.

A study in another Southern California community discovered that although more than 90% of the whites, African Americans, and Latinos had talked with whites about local issues and 83% of the African American residents had spoken with other African Americans, only one fourth of the Latinos had included fellow Latinos in their conversations about these subjects (Bockman & Hahn, 1971, p. 79). These data imply that efforts to foster the convergence of personal and political issues might stimulate growing levels of political activation particularly among Latinos. On the other hand, a survey of the members of gay and lesbian organizations in California reported that they were almost twice as likely to mention friends rather than family or parents as the most important factor influencing their political activity (Schockman & Koch, 1995, p. 19). No comparable data have been collected about persons with disabilities in Los Angeles or elsewhere.

In general, antipathy to minority groups has been articulated more successfully in controversies about relatively abstract subjects such as affirmative action or the denial of government benefits to undocumented residents than about topics involving direct and immediate interests. Efforts to translate policy debates into a discourse that focuses on individual or family problems thus may tend to alleviate racial and ethnic tensions in Los Angeles and to encourage the involvement of disadvantaged voters in the electoral process. In part, these changes also might entail attempts to modify the strict dichotomy between public and personal concerns that has inhibited the discussion of many local problems in the past. In some cases, social scientists have been able to demonstrate the linkage between major public policies and supposedly private matters. Wilson (1987), for example, has shown that the growth of single-parent households in urban ghettos is directly related to persistent unemployment that has reduced the "marriageable pool" of African American men. Yet relatively few researchers have sought to explore such connections, and politicians have been reluctant to talk about them. In a fundamental sense, therefore, plans to redefine major issues may depend on the success of proposals to restructure local political institutions.

An Empowering Assembly: New Proposals and Agendas

This analysis seems to buttress the pressing need to create new institutions to facilitate increased participation to reduce the inequities that have prevented politically significant portions of society from gaining representation in local government equivalent to their proportion in the population. The basic objective of this recommendation is to promote the inclusion of segments of the community that have been formerly excluded from the governance of Los Angeles. In many respects, the proposal might be considered a fourth-generation approach that seeks to move beyond the limitations of an exclusive reliance on the existing electoral process, geographical representation, and even cumulative voting as a strategy for empowerment. The principal purpose of this plan is to devise an empowering assembly of disadvantaged groups that would elect their own representatives to such a deliberative body. The members of this assembly would engage in communications designed to formulate positions on prominent issues that could be presented to their own constituents as well as to incumbent politicians and to the white or Anglo majority.

The initial phase of this endeavor might consist of a debate about the desirability of the proposal itself. These discussions would permit the groups not only to assess existing levels of agreement among themselves but also to begin the process of disseminating their viewpoints to a broader public. Through a series of open forum meetings and informal conversations, the self-selected leaders of disadvantaged groups would have an opportunity to participate in a dialogue about the merits of the idea and to decide whether or not they want to establish such an institution. If there is a general consensus to move forward after all aspects of this question have been considered carefully, these deliberations might culminate in a constituting convention to fix the details of the plan.

As preliminary criteria, I suggest that the assembly might include spokespersons from disadvantaged groups that have generally been unsuccessful in electing representatives to public office from districts composed primarily of voters in the dominant majority. Hence, representation may be extended primarily to groups that have formed significant social movements such as African Americans, Latinos, Asian Americans, Native Americans, women, gays and lesbians, as well as older and disabled persons. At a later stage, consideration could be given to selecting additional representatives on the basis of socioeconomic status, religion, or other factors. Initially, however, plans might be appropriately founded on an explicit recognition of the underrepresentation of disadvantaged portions of the population. Because the proposal is designed to create a forum for articulating collective, rather than individual, goals, each of the groups may be granted an equal number of seats in the assembly. Subsequent attention could even be devoted to the possibility of allowing more spokespersons to be chosen by the least powerful groups as measured by factors such as prior levels of representation, voting strength, or political financing. The right to become involved in this project also would be extended to members of disadvantaged groups throughout the Los Angeles metropolitan area. These recommendations are designed to be experimental and provisional; changes, of course, could always be instituted as further experience indicates.

The intent of this proposal is to create a mechanism whereby the perspectives and opinions of disadvantaged residents of Los Angeles can be made known to each other and to dominant segments of society. The recommendation purposely does not require formal authorization from existing officeholders, extensive changes in local ordinances, or major appropriations of government funds. The plan may encounter resistance from incumbent politicians who could perceive the emergence of new leadership as a potential threat to their jobs. But public officials would not be legally mandated to act on any of the suggestions advanced by the assembly. Instead, the goal is to generate enough media publicity and public pressure to compel the consideration of issues that might otherwise be neglected or ignored.

The opportunity to participate in the election of representatives to the empowering assembly could be made available to any eligible local resident who furnishes a minimal contribution such as a few postage stamps to cover the mailing of ballots and campaign statements. Additional funding for administrative expenses to start the program might be obtained from philanthropic foundations. To curb the influence of money on the decision-making process, persons seeking positions in the assembly would be prohibited from receiving or spending personal funds or private political donations. Candidates who qualify by circulating petitions would campaign primarily through public meetings and limited mailings. The plan is designed to give disadvantaged voters increased access to the political process at minimum cost.

Another major objective of the proposal is to foster the development of nonterritorial communities as an alternative to basing political representation solely on districts that tend to dilute the influence of continuing minorities. To begin this process, voters would be able to select the disadvantaged group with which they are most closely affiliated. Thus registration might be based primarily on self-identification. In view of the stigma that has often been attached to public identification with such groups, there seems to be little danger that citizens in the dominant majority might attempt to subvert this process by falsely claiming to be members of participating segments of the electorate. To prevent this distortion of democratic principles, a method of certifying eligible residents could be devised later, if it should prove necessary. Initially, however, the election of representatives to the empowering assembly would be designed simply to replace arbitrary geographical districting with procedures that reflect significant divisions within the community.

Members of the assembly would, of course, be free to discuss any matter that they choose to consider. The plan reflects the hope that increased dialogue between the leaders of the various groups would lead to the identification of new issues that might provide greater opportunities for unified action than topics that are ordinarily debated by local officeholders. Many of these concerns can be expected to revolve about personal and family problems including education, juvenile violence, drug abuse, and programs to support generational bonds. By engendering local interest and participation in innovative projects, measures that emerge from this assembly might eventually be placed on the agenda of city councils or other legislative bodies. In addition, representatives may focus on discriminatory policies and behavior that would otherwise receive little, if any, public attention. Perhaps most important, discussions in the proposed assembly would facilitate the formation of coalitions within and between disadvantaged groups. Clearly, spokespersons could achieve the greatest influence when these groups support each other or agree on a common position. As a result, the plan can be expected not only to increase the strength of underrepresented voters but also to reduce polarization by fostering alliances in a wide range of controversies through a process of negotiation, compromise, and conciliation.

A partial precedent for an empowering assembly is indicated by the function of local commissions that have occupied a crucial position in the government of Los Angeles for many years. The representation of African Americans, Latinos, Asian Americans, and women on the commissions has increased since Mayor Sam Yorty left office in 1973, but the appointees continue to be drawn primarily from elite occupations such as lawyers and business owners or executives (Gomez, 1994; Sonenshein, 1993, pp. 144-151). Moreover, these groups are divided between influential boards such as the commissions on the airport, the harbor, water and power, civil service, planning, the police, and public works and comparatively powerless bodies such as the commissions on aging, disability, human relations, Native Americans, and the status of women. Thus, the creation of the assembly would redress a severe imbalance in the importance attached to these subjects. Furthermore, the members of the new institutions would be elected, rather than appointed.

Adoption of the proposed plan, therefore, would introduce a democratizing element into a political system that has long been vulnerable to charges of elite domination.

This proposal is not intended to be a panacea for the vast difficulties that beset Los Angeles. No attempt has been made to address the fallacy of the conservative assumption that the need to restructure local institutions is obviated by the progress that disadvantaged groups have already made toward the fulfillment of their political goals. Nor has a direct response been provided to the major weakness to the plan, namely, the impracticality of expecting incumbent politicians to surrender significant prerogatives and resources voluntarily. Yet the strength of arguments concerning inequality and injustice cannot be considered irrelevant to the challenges that the metropolis must confront in coming years. Furthermore, although specific suggestions have been offered to form a framework for further discussion, it can be hoped that critics will not allow possible objections regarding these details to obscure a serious consideration of the merits of the idea. In an era that is devoid of easy remedies for persistent urban problems, perhaps the most—or the least— that can be done is to introduce a new perspective to the agenda of political debate. The formation of an empowering assembly constitutes a viable alternative to prior recommendations. The plan might not yield immediate solutions to the city's most pressing needs, but it could enliven public discourse about the future of Los Angeles.

References

Almaguer, T. (1994). *Racial fault lines: The historical origins of white supremacy in California.* Berkeley: University of California Press.

Altschuler, A. A. (1970). *Community control: The black demand for participation in large American cities.* New York: Pegasus.

Anspach, R. R. (1979). From stigma to identity politics: Political activism among the physically disabled and former mental patients. *Social Science and Medicine, 13A,* 765-773.

Bengtson, V. L., Cutler, N. E., Managen, D. J., & Marshall, V. W. (1985). Generations, cohorts, and relations between age groups. In R. B. Binstock & E. Shanas (Eds.), *Handbook of aging and the social sciences* (pp. 304-338). New York: Van Nostrand Reinhold.

Bobo, L., Zubrinsky, C. L., Johnson, J. H., Jr., & Oliver, M. L. (1994). Public opinion before and after a spring of discontent. In M. Baldassare (Ed.), *The Los Angeles riots: Lessons for the urban future* (pp. 103-133). Boulder, CO: Westview.

Bockman, S., & Hahn, H. (1971). Networks of information and influence in the community. In H. Hahn (Ed.), *People and politics in urban society* (pp. 71-94). Beverly Hills, CA: Sage.

Bond, J. M. (1936). *The Negro in Los Angeles.* Unpublished doctoral dissertation, University of Southern California, Los Angeles.

Carney, F. M. (1964). The decentralized politics of Los Angeles. *Annals of the American Academy of Political and Social Sciences, 353,* 107-121.

Davis, M. (1992). *City of quartz: Excavating the future in Los Angeles.* New York: Vintage.

DeGraaf, L. B. (1970). The city of black angels: Emergence of the Los Angeles ghetto, 1890-1930. *Pacific Historical Review, 39,* 323-352.

Dowd, J. J., & Bengtson, V. L. (1978). Aging in minority populations: An examination of the double jeopardy hypothesis. *Journal of Gerontology, 33,* 427-436.

Feagin, J. R., & Hahn, H. (1970). Riot-precipitating police practices: Attitudes in urban ghettos. *Phylon, 31,* 183-193.

Fogelson, R. M. (1967). *The fragmented metropolis: Los Angeles, 1850-1930.* Cambridge, MA: Harvard University Press.

Frederickson, G. (Ed.). (1973). *Neighborhood control in the 1970s: Politics, administration, and citizen participation.* New York: Chandler.

Gomez, J. (1994). *Minority power and city commissions.* Unpublished paper, University of Southern California, Department of Political Science, Los Angeles.

Grebler, L., Moore, J. W., & Guzman, R. C. (1970). *The Mexican-American people: The nation's second largest minority.* New York: Free Press.

Greenstone, J. D., & Peterson, P. E. (1968). Reformers, machines, and the War on Poverty. In J. Q. Wilson (Ed.), *City politics and public policy* (pp. 267-292). New York: John Wiley.

Greer, S. (1972). *The urbane view: Life and politics in metropolitan America.* New York: Oxford University Press.

Guerra, F. J. (1987). Ethnic officeholders in Los Angeles County. *Sociology and Social Research, 71,* 89-94.

Guinier, L. (1994). *The tyranny of the majority: Fundamental fairness and representative democracy.* New York: Free Press.

Hahn, H. (1968). Northern referenda on fair housing: The response of white voters. *Western Political Quarterly, 21,* 483-495.

Hahn, H. (1976). The American mayor: Retrospect and prospect. *Urban Affairs Quarterly, 9,* 276-288.

Hahn, H. (1992). The Los Angeles uprisings of 1992: A disability perspective. *Disability Studies Quarterly, 13,* 80-84.

Hahn, H. (1994). The civic uprising and intra-urban governance. In G. O. Totten III & H. E. Schockman (Eds.), *Community in crisis: The Korean American community after the Los Angeles civil unrest of April 1992.* Los Angeles: University of Southern California, Center for Multiethnic and Transnational Studies.

Hahn, H., & Holland, R. W. (1976). *American government: Minority rights versus majority rule.* New York: John Wiley.

Hahn, H., Klingman, D., & Pachon, H. (1976). Cleavages, coalitions and the black candidate: The Los Angeles mayoralty elections of 1969 and 1973. *Western Political Quarterly, 29,* 521-530.

Hobsbawm, E. J. (1965). *Primitive rebels: Studies in archaic forms of social movement in the 19th and 20th centuries.* New York: Norton.

Hofstadter, R. (1955). *The age of reform: From Bryan to FDR.* New York: Knopf.

Jackson, B. O., Gerber, E. R., & Cain, B. E. (1994). Coalitional prospects in a multi-racial society: African-American attitudes toward other minority groups. *Political Research Quarterly, 47,* 277-294.

Kinder, D. R., & Sears, D. O. (1981). Prejudice and politics: Symbolic racism versus racial threats to the good life. *Journal of Personality and Social Psychology, 40,* 414-431.

Kotler, M. (1969). *Neighborhood government: The local foundations of political life.* Indianapolis, IN: Bobbs-Merrill.

Minow, M. (1990). *Making all the difference: Inclusion, exclusion, and American law.* Ithaca, NY: Cornell University Press.

Moos, A. (1989). The grassroots in action: Gays and seniors capture the local state in West Hollywood, California. In J. Wolch & M. Dear (Eds.), *The power of geography: How territory shapes social life* (pp. 351-369). Boston: Unwin Hyman.

Mowry, G. E. (1951). *The California progressives.* Berkeley: University of California Press.

Oliver, M. L., & Johnson, J. H., Jr. (1984). Inter-ethnic conflict in an urban ghetto: The case of blacks and Latinos in Los Angeles. In R. E. Ratcliff & L. Kriesberg (Eds.), *Research in social movements, conflicts and change* (Vol. 6, pp. 57-94). Greenwich, CT: JAI.

Pettigrew, T. F. (1971). When a black candidate runs for mayor: Race and voting behavior. In H. Hahn (Ed.), *People and politics in urban society* (pp. 95-118). Beverly Hills, CA: Sage.

Preston, M. B. (1993). *The 1993 Los Angeles mayoral election: Partisanship in a nonpartisan city.* Paper presented at the Workshop on Race, Ethnicity, Representation, and Governance, Harvard University, Center for American Political Studies, Cambridge, MA.

Quarantelli, E. L., & Dynes, R. (1968). Looting in civil disorders: An index of social change. In L. H. Masotti & D. R. Bowen (Eds.), *Riots and rebellions: Civil violence in the urban community* (pp. 131-141). Beverly Hills, CA: Sage.

Rhode, D. L. (1989). *Justice and gender: Sex discrimination and the law.* Cambridge, MA: Harvard University Press.

Schockman, H. E., & Koch, N. S. (1995, March). *The continuing political incorporation of gays and lesbians in California: Attitudes, motivations and political development.* Paper presented at the annual meeting of the Western Political Science Association, Portland, OR.

Sears, D. O., & Kinder, D. R. (1971). Racial tensions and voting in Los Angeles. In W. Z. Hirsch (Ed.), *Los Angeles: Viability and prospects for metropolitan leadership* (pp. 51-88). New York: Praeger.

Skerry, P. (1993). *Mexican Americans: The ambivalent minority.* New York: Free Press.

Sonenshein, R. J. (1993). *Politics in black and white: Race and power in Los Angeles.* Princeton, NJ: Princeton University Press.

Wilson, W. J. (1987). *The truly disadvantaged: The inner city, the underclass, and public policy.* Chicago: University of Chicago Press.

Young, I. M. (1990). *Justice and the politics of difference.* Princeton, NJ: Princeton University Press.

Hollywood, U.S.A.

LEO BRAUDY
Professor of English, University of Southern California

One of the most pervasive stereotypes of Southern California has been constructed by film, with its emphasis on the Los Angeles of the eye, the ostentatious, self-advertising place in the sun that lured the rest of the country to it as a kind of paradise, a surging Eden where rebeginning was the norm, a spectacular combination of perfect weather, transcendental artifice, and bodily perfection.

But the Los Angeles of the written word, of fiction and poetry, has often been more critical and more jaundiced, less relentlessly idealized in its vision of Los Angeles, preoccupied with how the promise has turned so quickly into the loss. This is the Los Angeles of darkness and apocalypse. Its vision, conveyed in language rather than images, has also had a tremendous impact. It's almost impossible (for example) to talk about the look of a film such as *Blade Runner* and its many descendants without paying tribute to two kinds of writing whose roots are deep in Southern California: science fiction with its dream/nightmare of the future and detective fiction with its labyrinthine cities down whose streets the hero probes with little final success to find the answer to the ultimate mystery.

Lakewood, California 7
"Tomorrowland" at 40

ALIDA BRILL

> Utopia: An imaginary or hypothetical place or state of things considered to be perfect
> . . . a condition of ideal (especially social) perfection . . . an imaginary distant region
> or country . . . impossibly ideal, visionary. . . . Having no known location, existing
> nowhere.
>
> *New Shorter Oxford English Dictionary, 1993, p. 3534*

Towns and communities are just like people and institutions—they all have life cycles and
life spans. Some communities work better longer than others; some have robust and healthy
lives for a long time, developing new and richer communal histories and traditions; others
survive, even if happily, for relatively short periods; others adapt to changes and go on—although differently—continuously and even somewhat harmoniously. We are by now accus-
tomed to the usual litany of obituaries for the death of previously great cities. Suburban towns
in great distress, emotionally, economically, and communally, are less well known stories,
although they are not as uncommon as they might appear at first glance.

Sometimes, early visions of architects and planners, even when coupled with the
enthusiastic dreams, hard work, and loyalty of residents, run afoul of contemporary reality
and its demands and requirements for personal compromises. History can rarely be
predicted accurately enough to stave off all possible human disappointments. Perhaps
some town plans and housing designs are more adaptable than others to the inevitable,

AUTHOR'S NOTE: I am deeply grateful to Dr. Joann Vanek for her constant, valuable insights and helpful
recommendations. I am most indebted to the critical thinking and wisdom of Frances Madeson, who read several
previous drafts of this chapter.

Comments by Lakewood residents and officials in this chapter are from interviews conducted and conversations
held between February and June 1994.

albeit unknown, future shifts in demographics, as well as the changing patterns of social
and economic trends. Lewis Mumford certainly believed, as Jane Jacobs has concurred,
that some plans or designs may be more readily flexible and able to make required changes
in community structure for a healthy civic and personal life of a town, throughout the life
of the community itself. In his early book *Survival Through Design,* California architect
Richard Neutra (1954) detailed at some length a possible prescription to avert disaster,
especially in Southern California.

This chapter is a journey into a place that broke most of the rules Neutra and others
have outlined for community survival, in Southern California particularly—about community cohesiveness, self-reliance, architectural design, and town planning. Yet for
everyone involved at the time and for many years later, the place appeared a most viable
answer to the question, "What is a perfect family town?" The answer seemed to be a city
just south of Los Angeles called Lakewood. Although it never received the type of media
attention heaped on its "stepsister" on the East Coast, Levittown (New York), Lakewood
achieved its own recognition for its form of municipal government, a contract-based plan
known in the planning vernacular as the "Lakewood Plan." Beginning in 1993, Lakewood
became synonymous with troubles originating from a gang of neighborhood boys, the Spur
Posse, a story generally identified in print and broadcast media as the "sex for points scandal."
In four decades, or in little more than two generations, Lakewood has moved from an
enactment of the American dream, California style, to a town at war with itself, feeling it is
under siege from others. Life in Lakewood has deteriorated to the point where even its
next-door neighbor and "big sister" Long Beach is viewed as a potential adversary.

In retrospect, it was perhaps the deliberate notion of a family utopian town that has
caused part of the inability to adjust to current realities. Certainly, the planners believed,
in a most self-conscious way, that they were creating something of great value and unique
quality. At the time of its inception, the town was heralded as the first planned "garbage-
free" community, a reference to every home having an in-sink garbage disposal—a
curious and luxurious appliance in the early 1950s. What went wrong in Lakewood?
Nothing that different from what goes on in many previously middle-class, now barely
working-class, communities. But Lakewood provides a unique case study of the greater
Los Angeles area for a number of reasons. Mixed in with the usual gender displacement
of men out of work, fear of racial integration, change in the demographics of the schools,
and so forth is an entrenched mythology that presents itself as an almost indignant sense
of a loss of guaranteed entitlements. This sense of the loss of personal entitlement does
not revolve around those democratic guarantees of the Bill of Rights but instead involves
a special set of consumer and ideological communal rights residents forthrightly discuss.
The emotional and cultural ethos behind the Lakewood Plan seemed somehow to make
many original residents and their offspring (many of whom remain in Lakewood) believe
they were (and are) entitled to a lifetime of work, safety, homogeneity, and virtual
seclusion from the compromises other communities have had to make to remain viable.

For me, as a social critic and someone who writes about power and gender relations in
American society, Lakewood presents an obviously interesting puzzle with which to

work. But there is another far more personal reason that brought me to Lakewood in the early 1990s, just after the Spur Posse began to become national news. I am, as they say in Lakewood's parlance, an "original kid." I grew up in Lakewood, moved into one of the planned homes, and graduated with honors from the "jewel of the town," the eponymous high school, Lakewood High. To see the fantasy exposed in the 1990s as it has been is to digest both intellectually and personally what may have been its flaws from inception. The examination of the deterioration from the hopefulness of the first days to the fragmented and alienated community of today, with a goodly number of despairing and fearful residents, is not merely a scholarly exercise in the analysis of structural and gender dislocation. Obviously, it is that, but it is a personal odyssey as well.

A planned, post-World War II community, Lakewood was built with the hope and the promise inherent in the cultural ethos of the 1950s. A victorious spirit of optimism had gone into each hammer stroke in Lakewood's building. Louis Boyar, Mark Taper, and Ben Weingart formulated a plan, new and unique, that would make each of them rich and would alter forever the map of Southern California. They created a town of row after row of neatly sculpted individual lots, ready to become "front yards and backyards," with houses of the same shape, dimension, and scale. Despite the promotional rhetoric to the contrary about exterior trim styles and different floor plans, to an untrained eye they looked identical. Lakewood was the realization of a Lewis Mumford (1955) comment, an eerie warning when applied to mass-designed housing projects such as Lakewood: "The prospects of architecture are not divorced from the prospects of the community" (p. 88).

The successful completion of the developers' plan helped in the transformation of mass middle-class housing from its more primitive conceptual beginnings in the 1930s and 1940s to the reality of the 1950s. A town was created, and a map was made, where neither had been before. Acres of sugar beet and bean fields were turned into rows of housetops. The developers designed and manufactured a "dream" town that affected, stylized, and modified the lives of tens of thousands of people—an American dream best articulated by the idealized moms and dads of popular sitcoms of the time, such as *Father Knows Best* and *Ozzie and Harriet.* Coined "Tomorrow's City Today," Lakewood was a town deeply invested in the notion that it was *the* perfect family town, whose planners apparently had thought of everything—parks, pools, and adequate backyards. There were convenient shopping strips built into the plan for each few blocks of housing, and there was the unifying and enormous Lakewood Shopping Center. "The Center," as it was always called, broke all size records at the time for a shopping complex. Nearby churches of almost every denomination served the spiritual needs of the community, and "neighborhood schools" every few blocks provided a safe education for children, almost all of whom walked to and from school. If Lakewood were not the modern utopian community, it would have to do until the real thing came along. This ideology that Lakewood was the perfect place for families, especially children, is a belief still so deeply cherished that the contemporary reality is difficult to comprehend. Many residents simply refuse to move beyond the entrenched mythology, and many choose not to leave once the reality is confronted. With a mythology so powerful that it obfuscated the inevitable, Lakewood was the cultural accident waiting to happen.

By 1993, the dark side of the town's culture had become all too evident—what may have been a relatively private agony of a town in which many things had gone wrong for its residents became national news. Many moms were far from happy, too many dads were unemployed and angry, and the storybook children were running wild in the streets and on the high school campus. The group of neighborhood boys (all Caucasian)—popular and charismatic—had formed a club some years before, which became known nationally as the Spur Posse. To belong to the Spur Posse, the boys had to have as many sexual conquests as they possibly could "score"—each time with a different girl. The boys were then awarded points for each incidence of sexual activity with a new girl. From these numbers, which represented their weekly points, the correlation was made to the number of a famous athlete's jersey. Until the arrests were made (some of the girls were as young as 10), the boys thought it was all a game. Los Angeles District Attorney Gil Garcetti's office chose not to prosecute the boys, concluding that it did not have enough "ironclad evidence." Older Spur Posse boys were also known to harass and bully younger boys— ganging up on them at the parks, stealing their balls and bats, and beating them up. Something had gone drastically awry in the city of tomorrow.

For more than 3 years now, I have been trying through fieldwork and intensive interviewing to attempt to unscramble some of the causes of the decline. The outside journalists who came for a few days, or at most a few weeks, to write about the Spur Posse emphasized the "aerospace culture" of the place as a particular brand of factory town. Although it is true that for many years the aerospace and aircraft industry fueled the town's economy, the spirit of the place has never been "factory town" or "just like an old steel mill town." These outside journalists missed essential elements—which are hometown and home team, all the way. If Lakewood could be reduced to a single metaphor, *sports team* is much more accurate than *factory* or *aerospace*.

Lakewood is proud of its athletes. That pride and the concentration on sports as a way of life are in the air, in the organization of town leisure activities, and in the geography of the parks. The pride is most keenly felt in the continuing, and now somewhat empty, traditions of athletics at the troubled Lakewood High School. A number of watering holes such as the local Tuck's bar and the bars of the more interesting beach parts of Long Beach, such as Belmont Shores, still have evenings predominated by crowds of men who were the "great stars" of the Lakewood High "lean, mean, red, fighting machine." A local McDonald's houses the Lancer Sports Hall of Fame—a sports memorabilia museum paying tribute to the sports achievements and competitive triumphs of the town's high school athletes.[1] Although in recent years exhibits have included pictures of girls and testimony to their athletic prowess, the point of the McDonald's Sports Hall of Fame is the boys and the now men who were part of Coach John Ford's "glory machine" of the 1960s and 1970s.

More than 40 years after its incorporation as an independent city, the prongs of Lakewood's ethos, both cultural and economic, remain focused on values relevant in the 1950s—family and children, the safety of parks (especially on the availability of sports opportunities for children, beginning at an early age), and the expectations of a variety of middle-class entitlements. Perhaps most sad, Lakewood remains obsessed with its borders, hoping that it will somehow preserve its largely homogeneous, not racially or ethnically diverse, population of residents.

For many, the Spur Posse boys profoundly embarrassed a city that had already experienced the weakening of its masculine muscles. In a place in which feminism is usually called "women's lib" and feminists "damn libbers," and in which the phrase "bra burners" is as alive today as it was 25 years ago, it is not surprising that losing one's place is seen primarily as something happening to men. Men in Lakewood, because of many reasons, feel no longer able to act out their "heroic" masculine traits. Previously, they began this journey into the rites of male passage in Lakewood by excelling in sports at the parks at a young age, followed by high school sports, followed by either military service or a good job, often at the local McDonnell Douglas aircraft factory. Many men were reluctant to "trash" the Spur Posse boys because they themselves felt such dislocation and disjuncture in their own lives. For a society based so much on traditional notions of masculinity—war, sports, and work—few sanctioned opportunities for masculine "heroics" remain. Lakewood is even more complex for men because virtually every original dad had either fought in World War II or in Korea or had done his patriotic duty by staying home running the war machine in the aircraft factories producing fighter jets and bombs.

Joe Szabo, a Lakewood High School graduate now in his mid-30s, felt that although the Spur boys had misbehaved sexually, nothing better could be expected of them.

> This isn't a place for men anymore. Not Lakewood, not now. What is a man supposed to do here now? The shipyard is closed, they are killing the Naval base, there is no work at Douglas, no real military experience available in this country anymore. So, exactly what do you suggest a guy do in order to define himself? You want to know what is wrong with Lakewood today? I will tell you. What has gone so wrong with Lakewood is that men do not have a place they recognize as their own any longer.

The men of Lakewood are angry, unemployed or underemployed, and split apart by economic and racial tensions. The rise of the Spur Posse seemed to be an obvious outlet for an already damaged sense of self. Has Lakewood become a place in which there is so little left for men that they derive their sense of pride from the indiscriminate, criminal antics of their boys? There is also the usual panoply of complaints and worries about unemployment, health care costs, inflation, and taxes. Things both fundamental and basic are causing difficulty and sorrow. One afternoon when I stopped at a local coffee shop for a late lunch, I overheard in the booth behind me a conversation that seemed to highlight, in perhaps a homely way, the manner in which many residents view their world.

When the waitress served the food to the four people seated just behind me, one man said, with great anger, "Well, that about does it for this place. First they raise the damn price of the burger to $4.95 and then they stop giving you both French fries and salad. It just figures, it is the way the whole damn town is now." What followed was a heated discussion among the friends about the different places in town that had cut portion sizes and increased prices, often by just pennies. But it was the intensity of the feeling and the anger that was telling and that lingers in my mind. The town of tomorrow, in today's world, seems to be in the process of watching its dreams come unraveled and frayed since the hopeful days of its 1950s beginning. All the while, residents want to cling tenaciously to their attachment to a past that can never exist again. Lifelong resident Shirley Williams best articulated the mood of Lakewood's loyal: "You know, we are losing out here; we are losing our place, our parks, our schools, our town." The phrase "losing our place" was articulated by many residents in those exact terms and in other words that could be paraphrased in much the same way.

Perhaps because the vast majority of Lakewood's first residents were returning GIs, patriotism in Lakewood has revolved around honoring veterans and the war dead and is equated with the wars that have been fought by Lakewood's men and boys. Del Valle Park, historically a boy's hangout and, most recently, home base for the Spurs, has a used Marines' fighter plane at its entrance. In the 1950s, it was on the ground, and boys scrambled in and out of it. At some later juncture, it was elevated onto its current concrete winged bridges, making it appear as though in flight.

The Del Valle Park's memorial fighter plane has an "Indian Chief" decal on both sides of it (the old stereotypical depiction of a Native American Indian, a red-faced chief with two feathers in the back of his head). The plane retains its original military lettering, "Marines 17, 7L." Rows of red rosebushes have been planted at its base and roped off with a black chain link fence and white cement posts through which the chain link runs. On one side is a large plaque inscribed with the words "Dedicated to those who gave their lives in the cause of freedom in the Korean Conflict from 1950-53. Dedicated 1964." There are no names of those fallen in the Korean Conflict.

As Lakewood was dedicating a plaque to remember those killed in Korea, America was gearing up for a war that would take an even larger number of neighborhood boys to a foreign country. Some would come back from Vietnam; many would not.

The other side of the plaque remembers that war, and the fighter plane is now known as the Vietnam Memorial. Vietnam was disproportionately fought off the backs of African Americans, such as the boys from across town, graduates of a school known as Long Beach Polytechnic High, and off the backs of white boys from towns such as Lakewood. Towns in which the boys were not, and never would be, college bound. Standing in front of that memorial airplane and reading the names of those killed in Vietnam is like turning the pages of a huge family album. Lakewood is small enough and tightly knit enough that the names of the dead are familiar. Two boys from my neighborhood were killed, and a high school boyfriend was killed. Another, a Green Beret boy from the street behind my childhood home, came back with devastating psychological wounds apparent for years afterward. For Lakewood, Vietnam was a neighborhood war.

Besides listing the names of the Vietnam dead, the memorial plaque says "In remembrance of those of Lakewood who died in the service of their country in the Vietnam conflict." Underneath those words and names is another cast bronze plaque that is quite large. On it are inscribed the words of a poem written by a Lakewood Lancer, Dennis Lander—class of 1965, a Vietnam veteran. His poem is famous in Lakewood; he is remembered for it. What follows is an abridged version of his poem, "The Boys of Del Valle Park."

We climbed aboard that huge winged rocket
and rode it to the sky
Our minds would soar for hours and hours
We're never gonna die
With pitch and yaw, dives and rolls
We'd blast bad guys to heaven
We'd crash and burn and walk away
Heck we're only 7

But jets give way to bats and balls
To hoop and football too
"We've got great potential," they'd say
"The rest is up to you."
How quick time travels from innocent days
of running in the sun
Till the day your Daddy tells you,
"It's your duty, son."

"Be strong, be tough, be a man"
I'd listen to them say
"But wait a minute, everyone
I thought it was only play."
So Moms and Dads bring the kids on by
and read a name or two
Remember that these precious children
sure think the world of you

And think of us, if not out loud
But when it is quiet and dark
After all, we're your kids you know,
The Boys of Del Valle Park

Dennis Lander's poem summarizes boyhood and masculinity in Lakewood. Its sentiments make it easier to understand why so many want to sweep the Spur Posse scandal under the carpet. Part of Lakewood's mystique of masculinity has centered on sports and heroics on the battlefield, not openly aggressive sexual conquests. Approximately 20 years after the fall of Saigon, Del Valle Park, home of the war memorial, has become known in Lakewood as the unofficial headquarters of the Spur Posse.

The social space available for acting out masculinity was constricting in the context of a generally dim outlook for the economics of Lakewood, indeed all of Southern California. In many corners of the town, one heard quivers of worry—about the imminent closing of the naval base in nearby Long Beach and about the continued disastrous employment condition at the Douglas aircraft factory, the past mainstays of the region's economy. The reality of the diminishing local economy was made all too evident in 1994 and 1995 in the closed and barricaded Bullock's department store directly across from the civic center, awaiting demolition. When the Bullock's opened in the 1960s, it was celebrated as a civic event. Why would the opening of a department store in a shopping center have mattered so much? Because it was an upscale store, considerably classier than the older May Company, which had dominated the shopping center since its inception. Bullock's signaled something else; it signaled the town had arrived. It said Lakewood was still a boom town, continuing on its road of comfort and success. Bullock's grand opening gala featured school bands from the area and entertainment of all types. It was a festival.

In 1993 and 1994, a number of Bullock's stores all over Southern California were closed because of the hard economic times. (Indeed, most of them would not be able to survive the economic downturn in California.) Lakewood's had been closed for more than a year; in 1995, it was being demolished to make way for a megadiscount store more in keeping with the economics of present-day Lakewood. Driving by the front of the store, one could witness the machine that looks like a giant dinosaur doing its work—the same sort of machine that everyone saw on television in January of 1994, as it took bites out of the Los Angeles freeways that had collapsed in the earthquake. That same demolition dinosaur took a big bite out of the Bullock's sign, digesting the huge *O* as I drove past it one day. It seemed a too-perfect metaphor for what had happened in the 25-odd years since Lakewood toasted the opening of an elegant store and listened to the sounds of the award-winning Lakewood High School Lancer Band. The demolition of Bullock's was in a sad kind of synchronized rhythm with Lakewood's move away from a middle-class town to a blue-collar one—a move to a reality that the town was not eager to accept because Lakewood is a town holding on, trying to regain its place or to redefine that place.

For a long time, my hometown was known as "Lily White Lakewood" to our crosstown rivals, schools such as Wilson and Poly and Jordan—schools that had been integrated for years, some from their beginnings. Lakewood has been able to keep "its place" because others have previously stayed in "theirs." The Lakewood of the past depended on others "knowing their place"—the others including women, gays, Latinos, African Americans, and Asians. What "protected" Lakewood from intrusion by others was simply the slowness of change in economics and civil rights. Apparently, Lakewood residents believed something else—perhaps that they were in "Brigadoon" and could disappear, at will, into the mists to keep themselves safe from racial and ethnic diversity.

Former mayor Jackie Rynerson contends that Lakewood's very survival is linked to sustained racial and cultural integration. She steadfastly believes that the original residents will not feel overwhelmed if the pace is constant, but gradual. Many, however, already feel overwhelmed and frightened. Classmates of mine are shaken to see "so many blacks" in the shopping center.

> I am always on the lookout now. Things have changed here—at night, I mean; mostly, I am okay in the daytime; it is the night I worry about. I never have any windows open, day or night. My neighbor keeps her windows open, but I tell her that she is too lax. And we have this special security door installed now—did you ever expect to see that sort of thing in Lakewood?

Don Waldie, head of public information at the city hall, believes, like Jackie Rynerson, that the challenge of diversity must be met. He is a man of candor, but one with a positive commitment to and attitude about racial change in particular.

> We are in a transition to a diverse community. We will be more ethnically diverse; this community cannot be sustained if we continue as we are to be one group and one race. Lakewood will not continue as an all-white community; it cannot. Our survival depends on our ability to integrate with others. We have a history here. We have an enormous original population that is still here, a population which came from the border South and mostly, of course, the Midwest. The attitude was always, I'll work with you as a neighbor, and that is what built this community—neighbors working together. What you have along with that wonderful sense of community and neighborhood values that come from a border Southern and Midwestern culture, is prejudice. You do have prejudice.

The views of Waldie and Rynerson are not the predominant ones in town. Many see neither the inevitability nor desirability of diversity. One of those who most assuredly does not share that view is former mayor Larry Van Nostran. Van Nostran has conservative religious views about prayer in school and about the undesirability of integrating Lakewood and other views anathema to a liberal worldview or belief system. He fears a "take-over" of the town and has already experienced a sense of loss of his place and identity in Lakewood. It makes him profoundly sad and angry.[2] Much of the discussion about racial issues in Lakewood comes down to a discussion of the heavy use of the Lakewood parks by African Americans and Latinos, people who are seen as outsiders and identified as nonresidents. For a long time, it was virtually assumed that the parks were somehow private. Public parks for the private use of the Lakewood residents characterized the prevailing view. While still mayor, in 1993, Van Nostran recalled (in a conversation with me at his place of work, a new car showroom) the following personal experience:

> Last month, on the Fourth of July, I got home from work at about 6 p.m. I said to my wife that I had better get right over to Mayfair Park to be sure that everyone knew that for the first time in many years, there were not going to be any fireworks on the Fourth. I live right by the park, I got over there to tell everyone, make an announcement that there would not be fireworks. There was one white family, one white family—that was it. They were all by themselves, huddled by the pool. Everyone else was black, or they were Spanish, Latino, Mexicans, whatever. I saw this one Latin woman of some sort. She was very fat, a big woman, she had her large breasts hanging out. She was grilling hot dogs, her husband or boyfriend was giving her hickeys all over her neck. I walked all around that park and no one even knew who I was. After 19 years as an elected councilman in this city, those people did not even know who I was. I told my wife when I got back, well, what am I doing here? Why am I working so hard for this town, what is it all for now? Maybe it is just time to go.

Although not an expert in city planning, the superintendent of Long Beach Unified School District,[3] Dr. Carl Cohn, holds to his conviction that a contract city format would inevitably and ultimately become a problem for a place with Lakewood's value structure. Former mayor Van Nostran insists that the contract plan is the only viable form of governance for little cities. Whether Cohn is wrong and Van Nostran is right, or vice versa, is irrelevant when it comes to a discussion of Lakewood's park system. It is simply the case that no one who planned Lakewood's parks ever thought about them as anything other than neighborhood parks.

Most have no parking lots at all; Mayfair Park has a few spots for cars, and Hartwell, on the edge of town, has a decent-sized lot because almost all of Lakewood has to drive to get to that park. But the neighborhood parks that begin where the houses end, and end where the houses begin again, have no parking facilities whatever. In the Lakewood of the original 1950s mentality and design, kids walked to their own park or to the park where they were visiting a friend. The neighborhood park was their turf; the park they could walk to was their domain. When they went to another Lakewood park, they considered they were visiting. Bolivar Park is now a source of community tension. It is the park that is favored by minority people, especially African Americans. They are considered outsiders, and they are not welcome. An original resident said that they (the residents) "were in jail" on the weekends. "They are everywhere. We cannot leave our house, they park in front of our houses, we're trapped. We cannot leave, we cannot go out, and we cannot go to the parks."

I grew up as a Bolivar Park resident and learned to swim in the Pat McCormick Pool. That and the Billie Jean Moffitt (later King) tennis courts were two of the few signposts I had as a child that suggested I could achieve something outside Lakewood's borders. In point of fact, I preferred the Mayfair Park pool, but it was out of the question, except on special occasions or when visiting a girlfriend who lived near that park. Parks were places kids walked to and played at and swam at, unescorted by their parents. I recall that parks were also places where some boys were a bit rough and one had to pay attention. To pay attention primarily meant to avoid rude, not dangerous, behavior—at least not in the parks.

Former mayor Jackie Rynerson said the bitterness is complicated because people do not want to face the idea that some of the African Americans using the parks do in fact live in Lakewood. Some live in the Lakewood Manor Apartments, the only area of Lakewood to be integrated comprehensively. Lakewood Manor Apartments is a large complex of inexpensive, all-rental units built by one of the town's founders, Ben Weingart. Most in Lakewood refer to their distaste with the integration of the parks as the "Bolivar Park problem." A young woman named Chandra who works at the park became visibly upset when talking about what has happened there.

> I won't ever let them know that I am afraid of them, but I am. I am afraid that people like that are going to move in here. We had a gun shooting at Bolivar over a basketball game. That is scary to me. You should have seen the park this weekend—we had to go around for 3 hours with a picker to get the trash picked up. These people have no respect for our parks. They didn't give a shit about what they were doing. We gave them trash bags, but they refused to use them. They won't go to their own parks because they feel safer in Lakewood, at Bolivar, where they can run around with their families and not get shot at in a drive-by.

A Lakewood resident in her 40s, who was an original kid, had much more mixed feelings about the situation in the parks.

We do not have our parks any longer. You know that all of these parks were supposed to be for us, they were our parks. The only one we will go to now is Hartwell; sometimes we go there to walk. We can't go to ours [Bolivar] any longer. I don't resent the coloreds for coming into the park, though. They are mostly families, organizational groups, church socials, that sort of thing. They come in on the weekends and use the barbecues and the pool. They can't go to their own pools or parks because of the shootings. They want their kids to be safe; I don't blame them. But where are we supposed to go now? Where is the place for us? So, we all end up staying in our backyards and barbecuing in our own backyards, or whatever. Sometimes I have to ask myself, what are we supposed to do, in a community where you can no longer get near the park on the weekend. I have to ask myself that.

Jackie Rynerson views the concern about using the parks on the weekend as overblown.

Our true concern is that there must not be alcohol or drug use at our parks, there must not be. There must be rules at the parks, and they must be obeyed. That is a real issue and a legitimate concern, but the "we can't use our parks" refrain is a bit of a deception. The truth is that almost all Lakewood residents would rather be in their own backyards on the weekends, anyway.

The crowding of cars in front of residents' homes on weekends, caused by the lack of parking lots at the parks, is indeed a perplexing problem that even Jackie Rynerson and other more liberal residents admit to. Residents have been known to be trapped in their homes for entire days by cars parked in front of their driveways. As Rynerson said, "The fact is that while we considered these to be public parks, we did not, 45 years ago, contemplate that people from other cities would need to use the parks, to any great extent."

It is around the issue of the social integration of the parks that one feels the hopelessness and passivity that invade the spirit of Lakewood. Something is happening to the parks and to the residents. There is a general sense of being invaded, of being overwhelmed. There is no mentality present that suggests that any good could come from trying to integrate with the people using the parks. Instead, the options are to retreat to one's own backyard, literally and metaphorically.

Perhaps because all of Lakewood began at the same time, it has the feeling of a club—the feeling that because everyone started out together, residents are entitled to lifetime "charter membership." The feeling of beginning together, the reality of a town now filled with middle-aged, original kids who have stayed or come back, as well as the older original residents who have never left—all conspire together to create an insider and outsider caste system. Outsiders are so unwelcome in Lakewood, perhaps, because of its beginnings as the ideal hometown, however artificial the construction of that ethos. Helen Hansen, superintendent of the high schools for the Long Beach Unified School District and former principal at Lakewood, commented on the beginning of Lakewood and its correlation to the attitudes presently entrenched in Lakewood's culture.

Partially, I think Lakewood can be explained by the fact that the town itself all started at the same time. Everything was the same; everyone was the same. It did not develop like a normal city does, in layers, over time, with diversity always incorporated into the building of the city.

People saw Lakewood as pretty perfect; their aspirations had been met and so therefore they had their kids.

Don Waldie's words about the past and the future of Lakewood accurately target the strength and the inherent weaknesses in the town: the little city that had initially thought it had planned for everything, in a region that had believed for a long time that it was forever golden.

When people talk about the Lakewood of yesterday, or how wonderful Lakewood High used to be, it is a problem. The difficulty with nostalgia is that it doesn't pressure the person to make historical changes. It is important to be clear-eyed, along with being loyal. Both the physical and the social fabric of this place was well built, but you cannot project into it a myth of the past. You must not interject into that mythology of the past an obsession with the past, with the nostalgia of the way it used to be. I encourage people to have a genuine understanding of the city's history, but not to get lost in it.

If there is any place in the communal recollection of Lakewood that is part of the obsession with the past, it is the high school. The Lakewood High School of the sports "glory years"[4] is a memory people want to be lost inside. Memories of the Lakewood High School's past are the stuff of mythology and poignancy, at the very least. In 1969, the school was still almost completely white. It retained its character as a model neighborhood school. A reunion letter inviting the class of 1969 to its 25th reunion refers to the "stain on the memories" of Lakewood High, which means the busing that began in the early 1970s.[5]

There was such anger at the initial busing that some residents apparently did have meetings to try to find a way to "buy" Lakewood High School out of the Long Beach Unified School District. Lancer alum Joe Szabo's remarks characterize vividly the feelings of many.

Now it sucks. The high school sucks. These kids know it, the neighborhood kids, they know it, and they don't know it. They don't want to know it. What Lakewood High was, they know that, though, because they have heard about it all—the special classes, the football team, the marching band. They know it isn't the same; they know it is over. I'll tell you, you will find Lakewood is a really racist town now; really racist now. Sometimes I guess I sound like a racist too. But it's because of what they did over there, to Lakewood High.

The sense of the loss of one's place, for many, begins with the loss of Lakewood High as a neighborhood school. Residents refuse to accept the reality that nothing could have been done and instead hold to the conviction that the city council could have done something to "save the high school." Classmates, new residents, and original residents feel the high school was the "jewel of the town." As such, they feel that busing should not have been permitted in their town. Remarks such as "they took it away from us" and "you should see what they have done to the high school" abound. Some are more direct, less inhibited. Without blinking, a Lancer alumni couple said, "The niggers and the Mexicans ruined everything we loved, everything." Another Lancer from the class of 1964 refuses to ever drive by the high school, taking a quite extensive detour to get to his home to avoid the school.

> I will not go by that place and see what has happened, the fences, the gates, the signs about weapons, and then to see what pours out of that place, our place—no, I won't be put through that. I don't even want to see that place in my rear-view mirror.

On a Friday afternoon, I went over to the high school at the end of the school day. I found an unforgettable tableau in place. Neighborhood kids were piled up, clamoring to get out from behind the iron gates. (Lakewood High is a "locked down" facility with no entry or exit possible without permission and a school official locking or unlocking the gates. The high school is often referred to as "The Cage.") Three school administrators, wearing Lancer Red blazers, were waiting in front of the auditorium between the bused kids who were being loaded onto the 22 buses and the neighborhood kids waiting behind the bars.

These students have a generalized air of acceptance and resignation; they do not suffer from the pain of nostalgia that their parents do. For them, all the troubles simply amount to life lived in the present. In short, Lakewood High is "just the way it is now. All the other schools are locked down too. There is no question that it does look like a big cage, though." Not only is the school locked, it was totally fenced in about 2 years ago and resembles a correctional facility much more than a high school. Recently, the administration closed all the bathrooms except the ones in the gym. The problems with graffiti and with security in the rest rooms proved too daunting a task for the school officials. Only one set of bathrooms open in a school plant as large as Lakewood's with a student population of almost 4,000 is a source of constant complaint and bitterness.

I was in Lakewood the day Jacqueline Kennedy Onassis was buried. I met Sue Huber Miller and her daughter Cheri for lunch to talk about the changes in Lakewood. Sue and I had been together at Hoover Junior High when we were told that President Kennedy had been killed. We talked about that day in junior high, and then she said something about the death of Jackie Kennedy that illuminated much of the feeling in Lakewood as it turned 40. "Sometimes I think everything we cared about, everything that mattered to us is being taken away from us. This is what her death reminds me of, that they are taking everything away from us." Who were "they"? Probably forces she felt were bigger than she was, bigger than the stability of Lakewood's promise of a happy-ever-after life. For her, Jackie Kennedy's death was not just the final picture being pasted into the scrapbook of the Camelot era; it was more concrete evidence that Lakewood was losing out. The death of the former first lady was a sign. For her, it represented a further symbol of loss of her own past.

Our "glory years" were filled with groupings, stratification, pecking orders, and social castes, but no one thought about trying to change it or modify the system that made some insiders and some outsiders. It was accepted that some would be hurtfully excluded, whereas others would be included and would be powerful and in charge. It was presumably all right because we were all white, I guess. The kids living in the country club estate section formed a special status group; the football players and the fraternities and sororities, a different, and sometimes related, system of inclusion or exclusion, approval or disapproval. We pretty much all looked the same and came from essentially the same types of home environments, regardless of the cruel rigidity of the carefully crafted, if false, social distinctions. Today, neighborhood kids make no distinctions about where they live inside Lakewood's borders—a leveling of the town's playing field that began in the early 1980s. From Joe Szabo's days, the only distinction made was that if someone lived in the country club estates, "it would just be cool to hang out there, because they had a bigger house, that was it." For the current neighborhood students, what matters is if you come on the bus or not.

Compton, California, a few miles up the freeway, has been an essentially all-African American city for many years. People in Lakewood use the word *Compton* as a synonym for *takeover*. The word *Compton* has come to mean absorption by another culture. More than 20 years ago, the fear of "Comptonitis" was so severe when Lakewood High began to play Compton in the local league that Lakewood High School Principal Harold Judson insisted that he and the Compton principal shake hands on the field before the game started, in full view of the Lakewood fans. Countless numbers of times, one hears the apprehension voiced in the question, "Are we going to turn into Compton?" or "We will turn into Compton, if we don't watch out."

> I've got three blacks [families] on my own block, right now. Three now, and well, you know the problem with blacks, they have friends, and they have visitors. That is the problem. We can't encourage our people to stay if this keeps up. Our housing stock has stayed pretty solid, but some people can't be encouraged much more to stay.

Former mayor Van Nostran says he will stay in Lakewood for the present but will not commit to what he terms *eventually*. Whenever *eventually* might be, he was not sure. He says that he could no longer think of Lakewood as "an absolutely forever place," as he always had before.

> What is going to happen to Lakewood, we are changing, in a lot of ways. I used to go to Compton years ago, to the Sears, it was so nice. They called it the "model garden city" in those days. Now look at the place. It could happen here, to us. Lakewood could become another Compton if we are not careful.

Reverend Robert Bunnett, who leads a large Protestant congregation and who has a core value structure apparently in harmony with the majority of the town's residents, has serious worries about racial integration in Lakewood.

> We have a feeling of being encroached upon. I hear it all the time in the interfaith meetings we ministers have together, as well as from my congregation members and my neighbors. I would say that there are many people in Lakewood, not just in this church, who feel that we may be overwhelmed, and we are afraid.

I was in Lakewood the day of a partial eclipse, the last one to occur in this century. It was an odd experience because it was a cloudy eclipse and made things feel heavy, slow, and out of focus. The sky was a strange, smoky color. While I was walking across the parking lot at the shopping center, a woman walking next to me asked if I knew what was wrong with the "weather." "Something strange is going on." When I told her that it was an eclipse, she silently nodded and moved on. It was too correct metaphorically—an eclipse. The whole town appeared in eclipse—the economics, the racial attitudes, the chaos at the high school, and the tyranny of the Spur Posse. It reminded me of a quote from one of my favorite books, *Teaching a Stone to Talk* (Dillard, 1992):

> What you see in an eclipse is entirely different from what you know. . . . Usually it is a bit of a trick to keep your knowledge from blinding you. But during an eclipse it is easy. What you see is more convincing than any wild-eyed theory you may know. (p. 15)

One lifelong resident, another original kid, experiences deep and personal pain about the losses in Lakewood. Her life has a vulnerability to it. She, like a number of others, does not see many options left for Lakewood families such as hers. At times, a chilling and defeated acceptance washes over her.

> I watched a television program with psychics who foretold that everything was going to happen between now and the year 2000. And they said that really terrible things were going to happen, more earthquakes, everything bad. One of them said that she had no predictions past the year 2000. So maybe it will all end then; it is not that far away, you know. But that would be OK, as long as we would all be in heaven.

Perhaps the despairing millennialism and depression of my classmate who never left Lakewood are understandable, given the way the town has collectively chosen to resist the revolutionary social changes going on in its midst. Why Lakewood remains so stubbornly wedded to a vision that cannot be translated into reality and to a dream that once was, even at its inception, unrealistic, is part of the power and the drama of the town's zeitgeist. But perhaps it is a simpler explanation than social scientists, critics, or historians like to admit. If one can accept that in a sense Lakewood was, from its drawing board stage, a kind of utopian family community, perhaps the answer to the passive quality of so many (but clearly not all) residents is more easily understood. Lakewooders thought they had landed in utopia, where everything was to work perfectly—from the garbage disposals, to the marriages, to the kids, to the schools.

In the early days, the garbage disposal often jammed; reset buttons were worthless. Mothers and fathers became experts at unjamming them—usually with broomstick handles. When a broomstick handle was shoved down the disposal and given a good quick sharp turn, the disposal coughed up what was bothering it and went right back to work making the perfect little family click right along again in the "world's only garbage-free community." There are no magical broomstick handles to resolve Lakewood's current turmoil and sorrows. As a critic and as a former original kid, I see a warning about preplanned utopian communities of any type.

Notes

1. The entire Lakewood McDonald's franchise is owned by Lakewood High School graduate and lifelong Lakewood resident Tom Piazza. He remains a committed and loyal member of the community and believes in the cohesiveness of the sports focus in the town.

2. In keeping with the old adage that people are more complex than their stereotypes, however, Van Nostran staunchly defended the girls who had been victimized by the Spur Posse. He said to me that he "hated that the girls had been made to look like sluts and whores and the boys were just stars and studs. I knew this was going to happen, the boys would be heroes and the girls would be trampled. And I hate it, I just hate it." He is the person who first alerted one of the most famous feminist attorneys in Los Angeles to the plight of the girls.

3. Part of the problem for Lakewood is that the contract system, in a sense, also relates to the schools. Although called Lakewood High School, the school is actually part of the Long Beach Unified School District and as such is controlled from downtown Long Beach. That is the reason for the busing of children into Lakewood and one of the reasons for the bitterness of the residents. There are many apocryphal stories about whether the schools inside Lakewood might have been able to have been purchased out of the Long Beach system. So, in fact, the very notion of "neighborhood schools" in Lakewood was a mythology from the beginning. It is hard to have neighborhood schools if the community itself has no control over what really happens in the schools within its borders.

4. For slightly more than a decade, the victorious Lancer football team, coached by the tough and legendary John Ford, was the defining center of the town's ethos. The city of Lakewood, so proud of the high school's sports record in football, unofficially adopted the high school's colors of red and white as the town's colors.

5. Excerpts from the reunion letter inviting the Lancer class of 1969 to its 25th reunion party:

Fortunately, for those of us in the graduating class of 1969 we remember a very special time of life, that is fast becoming extinct these days. No it wasn't Mayberry, but it certainly was a "kinder, gentler" time to head to class. . . . It was a time unstained by the violent nonsense stalking school hallways today. . . . It was a time when r-e-s-p-e-c-t wasn't just a Motown hit. It was a time when gay still meant bright and lively and "butch" was still just a haircut. It was a time when most of the "scoring" happened on the football field and not racked up by some posse. . . . It was a time when our school was not fenced in. Ah, but that was then.

References

Dillard, A. (1992). *Teaching a stone to talk*. New York: HarperPerennial.
Mumford, L. (1955). *Sticks and stones: A study of American architecture and civilization*. New York: Dover.
Neutra, R. (1954). *Survival through design*. New York: Deutsche Ausgabe.
New shorter Oxford English dictionary on historical principles. (1993). Oxford, UK: Clarendon.

Folio: Bordergraphies

Huellas Fronterizas: Retranslating the Urban Text in Los Angeles and Tijuana

ADOBE LA

Architects, Artists, and Designers
Opening the Border Edge of Los Angeles

Folio: Bordergraphies
Huellas Fronterizas: Retranslating the Urban Text in Los Angeles and Tijuana*

ADOBE LA
Architects, Artists, and Designers Opening the Border Edge of Los Angeles

Bordergraphies/Huellas Fronterizas: Retranslating the Urban Text in Los Angeles and Tijuana is a continuing urban research projected conducted by ADOBE LA. The research is focused on two cities—Los Angeles and Tijuana—to reconstruct and document the past, present, and future trajectories of cultural (inter)changes at the border and their relationship to migration and sociocultural change in the Mexican community in Los Angeles.

This multidisciplinary and comparative approach to the subject will be examined through the social, political, and economic conditions of urbanization. A primary concern of the research will be the effect of politics and visual representation in popular culture, art, and use of physical space.

*Multidisciplinary research project funded by the Fideicomiso para la Cultura México/USA—USA/Mexico Fund for Culture (Bancomer Cultural Foundation, the Rockefeller Foundation, and Mexico's National Fund for Culture and the Arts).

Tijuana Images

Photo No.	Title	Category	Description
1	The Border Line	Border Coded Messages	Bilingual plaque and jurisdictional territorial line between Mexico and the United States, Treaty of Guadalupe Hidalgo, 1848
2	Going North	Border Coded Messages	Pedestrian sign "tagged" at the San Ysidro border checkpoint
3	Crossing Borders: Sign of the Times	Border Coded Messages	Freeway sign at the San Diego-Tijuana border
4	(De)Fence	Symbolic Markers	Border fence made from scrap metal panels from the Iraq-U.S. war
5	Backyard	Vernacular INSurgent Urbanism	Border fence used as a back wall of house at the border line
6	Between a Cheeseburger and La Virgen	Border Urban Props: Myths, Masks, and Stereotypes	An urban carpet unfolds kitsch ceramics to welcome tourists that show, reinforce, and/or contradict stereotypes about Mexicans and North Americans
7	Bart Sanchez	Border Urban Props: Myths, Masks, and Stereotypes	Fluorescent Aztec calendars and Virgens de Guadalupe next to Mexicanized versions of Bart Simpson and the Power Rangers
8	CTM: Confederación de Trabajadores Mexicanos	Vernacular Graphic Design	Graphic sign for a shoe repair shop that belongs to the National Workers Union (CTM), the oldest and largest workers' union in Mexico
9	Border (De)Constructions	Vernacular INSurgent Urbanism	Hill houses made from scrap building materials from surrounding maquiladoras and factories near the border
10	Room With a View	Vernacular INSurgent Urbanism	Hotel, strategically located near the border fence and whose circulation and public areas are facing the Mexican-U.S. border

Tijuana Images, Continued

Photo No.	Title	Category	Description
11	La Mona (The Doll)	Vernacular INSurgent Urbanism	Five-story house built in the shape of a nude woman
12	Altar a Colosio	Popular Border Monuments	Neighborhood altar in memory of assassinated Mexican presidential candidate Donaldo Colosio
13	Plaza a Colosio	Popular Border Monuments	Government sponsored monument and plaza to Donaldo Colosio, built in the same site as the community altar

Photo Credits

Ulises Diaz: 13
Julie Easton: 1, 2, 4, 5, 6, 7, 8, 9, 11
Gustavo Leclerc: 10
Alessandra Moctezuma: 3, 12

Acknowledgments

ADOBE LA would like to thank Fideicomiso para la Cultura Mexico/USA—Fundación Rockefeller, Fundación Cultural Bancomer, and Fondo Nacional para la Cultura y las Artes for their generous support and funding of *Bordergraphies/Huellas Fronterizas: Retranslating the Urban Text in Los Angeles and Tijuana*. We also thank Mike Davis for his invaluable friendship and continuing support; Ignacio Fernandez for his friendship and eternal support on our journeys; the crew at Southern California Institute of Architecture (SCI-Arc); Michael Rotundi, Margaret Crawford, and Margie Reeves for their encouragement and support; SCI-Arc students Nancy Lopez and Kyung Jin Lee for crossing borders with us; Tijuana artist Marcos Ramirez "erre" for showing us the other, "hidden" Tijuana; sculptor Armando Muñoz Garcia for sharing his ideas behind his live-in sculpture, *La Mona*; Henry Jaramillo for his graphic design sensibility and patience; and Julie Easton and Barbara Jones for their photographs. We would also like to thank Holly Harper and Raul Villa for their insight and generosity of spirit. And finally, thanks to our families and friends—the foundation for this all. We are deeply grateful to them.

1. The Border Line

2. Going North

3. Crossing Borders: Sign of the Times

4. (De)Fence

5. Backyard

6. Between a Cheeseburger and La Virgen

7. Bart Sanchez

8. CTM: Confederación de Trabajadores

9. Border (De)Constructions   10. Room With a View

11. *La Mona* (The Doll)

12. Altar a Colosio

13. Plaza a Colosio

A Small Introduction to the "G" Funk Era

Gangsta Rap and Black Masculinity in Contemporary Los Angeles

TODD BOYD

Introduction

In a relatively short time, American popular culture has witnessed a radical departure from Bill Cosby's prominent image of Black masculinity that seemed to define the 1980s. In less than 10 years, the most visible image of African American maleness that was previously defined by wealth, status, and what many would call an overall "positive" image has been changed to highlight a oppositional image that exists in poverty, remains marginal, and characterizes what many have often called "the scum of the earth," in a colloquial sense, yet that fits more closely with the idea of "the truly disadvantaged." I am specifically referring to the image of the gangsta and his embodiment in the cultural medium of gangsta rap.

This chapter explores the cultural context that defines gangsta rap, focusing on historical significance, the furthering of the oral tradition, the cultural currency of Blackness, the primacy of visual imagery, the mediated aesthetic known as the hyperreal, and the way in which these issues come together around the larger dilemma of Black masculinity and its relationship to the contested space of modern-day Los Angeles.[1] This discussion also addresses the differences between African American folk culture and popular culture, the resistant anti-authoritarianism of gangsta rap that has even involved the FBI, and the

127

declining significance of overtly political discourse in rap that has come about in the era of gangsta rap.

It was some time ago that Gladys Knight suggested that "L.A. proved too much for the man" (in "Midnight Train to Georgia"), and although she was talking about a fictional character who had tired of the urban sprawl of Los Angeles and longed for the peace of a mythic Georgia, her appellation works quite well with the contemporary phenomenon of gangsta rap. With the storied past of Los Angeles informing a great deal of Black popular culture, be it the Hughes brothers' use of the Watts riots to begin *Menace II Society* (1993) or Dr. Dre's use of various voices from the civil unrest of 1992 on the track "The Day the Niggaz Took Over" (1993), it is necessary to analyze the function of this most intriguing city and its overall influence on the cultural artifacts that emerge from its domain.

Gangsta Rap: From the Marginal to the Popular

The explicit declamatory lines of "Gangsta, Gangsta" from the album *Straight Outta Compton* (1988) demonstrate the dominant thematic sentiment of defiant lower-class Black ultramasculinity that has circulated throughout the genre of gangsta rap. This album by Niggaz Wit Attitude (N.W.A.)—a group consisting of Ice Cube, Easy E, Dr. Dre, MC Ren, and DJ Yella—represents the inaugural blueprint on a public level for material that would eventually function beyond traditional musical or generic conventions.

The initial underground popularity of this music, set in place by N.W.A., would come to situate an entire cultural movement. As evidence of the group's societal impact, even at this early stage, the FBI targeted N.W.A. and their song "Fuck tha Police" from this same album as subversive and threatening. This gesture furthered the masculine mystique and allure of gangsta rap in a genre that relies on embellishing the menacing quality so often associated with the urban *lumpenproletariat*. What better way to demonstrate one's position as a true "menace to society" than by having one's lyrics cited as threatening by the highest echelon of government law enforcement?

In the parlance of the street, men are praised, with phallic connotations, for their ability to be "hard"; this is a modern-day variation on the sentiments that used to be referred to as "cool." To be hard, one must exist at all times in a state of detached utter defiance, regardless of the situation. By clearly enunciating the overt nature of societal defiance as communicated through "Gangsta, Gangsta," these lyrics reference a direct trajectory from the earliest days of African American folklore as expressed through oral culture. There has always been a proverbial "bad nigga" who, in serving notice on his presence, was feared by all around him—White people as well as people in his own community. Whether characterized by the images of Jack Johnson, Bigger Thomas, or the legendary Dolemite, it is apparent that rap music, especially gangsta rap, has successfully drawn on this tradition and rearticulated it for a contemporary audience.[2] Following in the tradition of the "bad nigga," the FBI's misguided gesture in the case of N.W.A. inadvertently helped sustain the defiant posture of masculine aggression that constituted a substantial portion of the thematic foundation on which gangsta rap would evolve.

At the conclusion of Bill Duke's *Deep Cover* (1992), Dr. Dre and Snoop Doggy Dogg in their rap tune of the same name created a direct link between gangsta rap and the medium of cinema. In the same way that Black music has always informed cinematic representations of African American culture, this historical juncture foregrounds the point at which gangsta culture would begin moving from a position—not unlike folk culture—to a much more public place in American popular culture. "Deep Cover" also shows the way in which gangsta rap culture is highly influenced by Hollywood images of excess, and, in turn, Hollywood is able to appropriate and merchandise the rather luminous visual imagery conveyed through the oral themes of gangsta rap.

In emphasizing this transition from folk culture to popular culture and the importance of the visual in this equation, it is important to single out Dr. Dre and his partner in crime, Snoop Doggy Dogg. They have been primary catalysts in mainstreaming gangsta culture throughout contemporary society. From late 1992 until well into the fall of 1993, Dr. Dre spent close to a year with a Top Five album, which went multiplatinum in sales, and had three videos in regular rotation on MTV. It is quite obvious that Dr. Dre represents a radical departure from the days some 2 to 3 years earlier, when Ice Cube's first solo effort, *Amerikkka's Most Wanted* (1990), went gold in 5 days with no radio or video airplay.

Although Ice Cube's unparalleled success, relative to the dictates of contemporary marketing strategies in the music business, suggests an almost utopian underground circuit of African American listenership, Dr. Dre's and, subsequently, Snoop's popularity in mainstream society represents the way in which gangsta culture has grown into a pop cultural force with few competitors. It also demonstrates the way in which Blackness, especially in the form of hardened Black masculinity, has become a significant commodity in the currency of contemporary popular culture.

In the past, this popularity would have been a sure sign of accommodation because the music would have to be compromised in some major way to be made mainstream. I argue that gangsta rap has come to prominence because of its unwillingness to compromise in a popular market that encourages accommodation. The music and culture industries have found ways to sell this extreme nonconformity, and many rappers have successfully packaged their mediated rage for a mass audience. Many audience members of all races use the music as a form of resistance or rebellion, be it adolescent angst, minority disenfranchisement, or an overall societal cynicism. With the truly disadvantaged male Black serving as supreme representative of angst, disenfranchisement, and cynicism in American society, gangsta rap has the ability to provide a vehicle for cathartic expressions across a variety of subject positions beyond a normally assumed exclusively Black space.

Coming on the heels of the subversive mystique generated by the FBI inquisition, Dr. Dre went on to demonstrate not only that this masculine defiance could threaten virtually every aspect of society but, more important, that this otherwise menacing form could cross over to mass audiences in a most popular fashion. Although hardened images of defiant masculinity have always had a strong appeal in American culture—for example, John

Wayne and Edward G. Robinson—seldom, if ever, has this sort of imagery been associated with Black men.

Although Black men have always functioned as the epitome of masculine subversiveness in regard to the larger society, seldom has this ethos, which emanates from a specifically lower-class Black male perspective, sold an extremely large volume of product without making thematic compromises along the way. The extreme nature of their performance substantiates the value of this cultural form for the rappers and the audience members who subscribe to the music. The mainstreaming of this type of unadulterated Blackness as cultural currency, however, is often sold to people whose racial and social circumstances are the complete opposite of those producing the culture. This recent situation allows the most marginal aspects of African American folk culture to come to financial fruition, albeit primarily for the White-owned music companies, in the form of gangsta rap. This financial profit also allows for the continued proliferation of a truly disadvantaged perspective via the mainstream media.

> I had no idea of peace and tranquillity. From my earliest recollections there has been struggle, strife, and the ubiquity of violence. . . . I've always felt like a temporary guest everywhere I've been, all of my life, and, truly, I've never been comfortable. Motion has been my closest companion, from room to room, house to house, street to street, neighborhood to neighborhood, school to school, jail to jail, cell to cell—from one man-made hell to another. So I didn't care one way or another about living or dying—and I cared less than that about killing someone. (Shakur, 1993b, p. 103)

Clearly blurring the lines between fact and fiction is this statement made by former gang member Monster Kody Scott (also known as Sanyika Shakur) of the notorious Eight Tray Gangsta Crips of South Central Los Angeles. What Ice Cube, Dr. Dre, and Snoop have fictionalized through their highly stylized narratives of gang culture, Monster has rendered in autobiography, thus linking these contemporary concerns to the historical by using a form that embodies such relevant tradition for the overall understanding of African American culture. The publication of Monster's memoirs, *Monster: The Autobiography of an L.A. Gang Member* (Shakur, 1993b), truly represented the solidification of the gangsta genre as a cultural movement in contemporary American society. This movement had now evolved from the marginal confines of regional association, through the heights of media-generated popularity, to an assumed cultural sophistication that could be conferred only by the substantiation of this lifestyle in literary form. The gangsta genre had now clearly moved into a multimediated arena that defied charges of trendiness or simple exploitation. The movement had become an integral part of both African American culture and American popular culture in a most effective fashion.

As one of the most sought-after items on the 1992-1993 literary circuit, Monster details the intricate nuances of gang life in South Central Los Angeles from the mid-1970s to the present. In addition, the impending publication of the text was the subject of extended features in *Esquire* (Shakur, 1993a) and the *Los Angeles Times Magazine* (Wallace, 1993) several months prior to the release of its hardback edition. Amy Wallace's question in the introduction to the *L.A. Times* piece, "How did an Eight-Tray Gangster Crip named

Monster go from hoodlum to literary hot property?" (p. 16), suggests the quest toward understanding the culture industry's embrace of gangsta iconography and the societal fascination associated with this most recent version of gangsters in American society. The publication of Monster's memoirs demonstrated that in addition to its public popularity, this area of culture should be regarded as an important commentary on the state of contemporary urban postindustrial existence for lower-class African American men. The literary reification of gangsta culture through Monster Kody's autobiography forced mainstream society once again to take "seriously" those voices that operate on the margins of society.

Monster's statement represents the point at which the image of the gangsta had evolved to a cultural movement that would include music, film, video, and, most important, the domain of literature as a legitimating strategy consistent with societal notions of artistic and political credibility. Consistent with the notion of a cultural movement are the repeated concerns that define my interest with gangsta rap culture: the issue of Black masculinity, both in race and class; the proliferation of oral culture; and the importance of understanding Blackness as a commodity.

These conditions are taken to another level at which the dictates of corporate entertainment culture serve as a backdrop to the massive proliferation of these images via mainstream channels of distribution. Many suggest that mainstream society has once again exploited Black culture for its own purposes—in this case, using the image of the gangsta to stereotype male Blacks as pathologically aggressive while laughing all the way to the bank. Yet this sentiment fails to incorporate the changes that have occurred surrounding the cultural production of "Blackness" in the post-civil rights era, especially as they relate to the truly disadvantaged position.

Cash Rules Everything Around Me

Contemporary society has allowed the limited participation of African Americans in mainstream culture as long as it continues to be profitable to the corporate interests that provide structural support for the presentation of cultural artifacts. Nevertheless, many African American performers and producers openly embrace this most recent form of "exploitation" as long as it provides them with the material possessions that comfort them in an otherwise uncomfortably racist world.

Gangsta rap, as it is fully informed by themes of excess, takes the notion of "gettin' paid" or "gettin' mine" to its most extreme form. Dr. Dre provides an example of this mentality in "The Day the Niggaz Took Over" (1993). In a previous generation, the tune's title would have been a declaration of nationalistic intent; here, "taking over" has to do with acquiring material possessions by any means necessary. Dre begins the tune by re-creating the atmosphere of the L.A. riots of April 1992 through the angry voices of those who were on the streets during this occurrence. Then he declares his own disposition regarding the true politics of the riot, "sittin in my living room/calm and collected/feeling that gotta-get-mine perspective." Thus, here, the riots are not defined as a political reaction to the ridiculous verdicts of the Rodney King case; instead, they become a venue for the excessive embrace of material possessions through looting. This is furthered by Snoop's rhetorical refrain, "how many niggaz are ready to loot?"

This cultural situation is quite similar to Cornel West's (1993) societal argument regarding the slow destruction of the Black community. West states that a great deal of decline can be attributed to "fortuitous and fleeting moments preoccupied with getting over—with acquiring pleasure, property, and power by any means necessary" (p. 5). He goes on to define the contemporary cultural climate as "a market culture dominated by gangster mentalities and self-destructive wantonness," which pervades all Black society, but "its impact on the disadvantaged is devastating, resulting in the violence of everyday life" (p. 14). Clearly, the politics of gangsterism has come to replace the strong nationalist sentiment that defined previous politics endeavors. According to West, this gangsta ethos functions as "the lived experience of coping with a life of horrifying meaninglessness, hopelessness, and lovelessness. The frightening result is a numbing detachment from others and a self-destructive disposition toward the world" (p. 14).

This nihilistic sentiment is once again echoed as Snoop declares in the opening to Dr. Dre's *The Chronic* (1992) that gangsta culture is represented by "niggas wit big dicks, AK's, and 187 skills." Excess and the overt symbols of oppressed Black masculinity are expressed through the perversion of an exaggerated phallus, high-powered weaponry, and the ability to kill at will. This perversion does much more than simply fulfill the societal stereotypes of the threatening male Black. Instead, it denies the White supremacist denigration embodied in the stereotype and reverses the impact to become a true purveyor of unadulterated Black rage.

Yet as the underside of this perversity continually leads to the obvious marginalization of Black women, it also has some connection to the continued extermination of lower-class Black men. Thus, it is necessary to acknowledge that we have truly seen the death of politics and the emergence of nihilism as the ruling order of the day. It is this sentiment that defines much of gangsta rap, certainly as it has entered the domain of mainstream culture. Although this is clearly the most obvious shortcoming of the genre itself, we cannot ignore the extent to which excess informs gangsta rap and culture, both thematically and in overall image. Acknowledging these societal tragedies is important, but they are concerns that constitute a large portion of public dialogue about race, class, and masculinity. To shift focus to the thematic of excess, relative to the accumulation of capital and commodity, and on an overall personal level, makes for a vivid cultural palette and allows a discussion of the cultural elements of this situation.

In one view, the excessive embrace of capital, sexism, violence, and an overall nihilistic mentality can be seen as the strident pathologies of gangsta culture. Yet I encourage a critical distance that moves away from traditional analysis, which seeks simply to expose the intricacies of each pathology. Instead, my focus is on the way in which these issues reverberate within postmodern American culture, specific to the African American experience. Once again, these social issues cannot be ignored, yet the focal point becomes that of representation, especially as it is configured in a visual sense, and the way in which the question of excess or spectacle redefines the understanding of these issues in contemporary society.

Although I am careful not to minimize the impact of racism or fall into a redundant posture of "blaming the victim," I am suggesting that we begin to focus our attention on the issue of racial representation in a postmodern society and the way in which this affects the incumbent articulation of Blackness throughout popular culture. I am advocating a critical distance away from the moralistic dimensions of African American cultural criticism, such as those put forward concerning positive/negative images; instead, I am arguing for a critical disposition that looks to highlight capital, commodity, and the way in which personal responsibility plays into the overall understanding of gangsta culture.

The Visual, With Los Angeles as Metaphor

The representation of gangsta culture is now prevalent throughout America, a far cry from the days when it was only heard at local swap meets and sold from the trunks of cars—when the music was present only in the various "hoods" of South Central Los Angeles, Watts, Compton, Long Beach, and Inglewood. This situation is underscored most effectively when referencing an actual component of the culture itself. DJ Quik's (1992) single "Just Lyke Compton" is a comical reflection on the impact that gangsta rap and its presentation through mainstream media have had on various sectors of American society that one would easily have considered quite immune to influences that derive from the regional confines of South Central Los Angeles.

Through a series of geographical references (Oakland, St. Louis, San Antonio, and Denver), DJ Quik describes each city's response to the spectacle of mediated gangsta culture. Quik details the way in which he encountered several imitation gangstas during his concert tour who embraced the media-generated imagery of the culture without the history or personal experience that inevitably makes one a participant in the culture. Quik suggests that the impact of gangsta culture, most effectively demonstrated through references to popular films, has misled the minds of those who use this imagery as a model for ultramasculinity, making associations with gang-specific areas of Los Angeles (Rolling 60's/ Crenshaw) although they have never inhabited these public spaces. It is as if the media presentation of these areas on a continual basis is a substitute for the actual experience of having been there.

In other words, Compton, a convenient metaphor for the entirety of gangsta culture, came to influence large sectors of society, although few knew the specific circumstances that created this aura around Compton in the first place. Supported by N.W.A.'s evolutionary *Straight Outta Compton* (1988), Quik goes on to ask rhetorically, "How could a bunch of niggaz in a town like this have such a big influence on niggaz so far away?" The perpetrators, having blurred the lines between the mediated life of a surreal Compton and their own real lives, have been victimized by the suturing effects of mainstream representation. Real lives have been lost as a result of the violent connotations associated with a ubiquitous image of "Compton," which in actuality is quite small compared with the vast landscape of Los Angeles.

By calling attention to the media spectacle surrounding the metaphoric image of Compton, Quik argues for authenticity and a historically specific, geographically based understanding of gangsta culture and critiques the massive proliferation of this imagery by those without the requisite knowledge base. Yet as time passed, mainstream media representation made DJ Quik's concerns about as comical as the recurrent themes in his song.

Much like the argument rendered by DJ Quik, this chapter explores the historically specific nuances of gangsta culture. The representation of the "gangsta," or "G," within popular culture is the context within which gangsta rap flourished and is specific to the late 1980s and early 1990s. The cultural movement is also geographically specific to Los Angeles and the West Coast in general. We have witnessed a proliferation of what I call the "articulation of Black Los Angeles": The musing of gangsta culture fits neatly into a developing pattern that also includes the films of Charles Burnett; the gritty revisitation of the hard-boiled, noir literary genre from a Black perspective specific to writer Walter Mosely; and critical reflections on being Black in Los Angeles by Lynell George (1992) in *No Crystal Stair*.

This movement surrounding gangsta culture takes its representational cues from 1970s African American cultural production and thematically references the often complex and contradictory nature of urban, lower-class Black male existence in a racist and nihilistic society. This culture is simultaneously composed of fictional as well as real-life occurrences, mediated as well as nonmediated events. It is both real and imaginary, often at the same time. This culture involves the mediums of music, television, film, video, literature, and the lives of everyday people. It is, in a word, postmodern: a series of events that could have happened only when they did specific to the societal currents that define our existence.

Ain't No Future in Yo Frontin':
The Issue of Cultural Authenticity

Another side of the argument exposed through the lyrics of DJ Quik supports the idea that the visual has become prevalent as a way of understanding the reproduction of African American culture. I am reminded of a *Wall Street Journal* front-page story titled "How a Nice Girl Evolved Into Boss, the Gangster Rapper" (Pulley, 1994). In the piece, reporter Brett Pulley contrasts the real-life background of Lichelle Laws with her gangsta rap image known as Boss. He details her attendance in Catholic school, along with her study of ballet, modern dance, and the piano—all signs of a middle-class upbringing. Yet these episodes are contrasted with her newfound persona of Boss, one of the first and most successful female gangsta rappers in a genre dominated not only by men but by a misogynist undercurrent. Pulley sees the two sides of life as incompatible; one cannot simultaneously be a middle-class woman and a gangsta.

Pulley fails to understand, however, in line with the overall infatuation with visual spectacle, that image, whether real or fictional, is the defining characteristic of contemporary society, especially in gangsta culture. In a sense, the image of Boss the rapper, a hardened violent woman on a murderous warpath, obliterates the real-life circumstances that may have defined her existence previously. It is not as though Boss conceals this luxurious past, because her mother's voice is heard lamenting over the loss of her "little girl" from the answering machine on the opening track of her debut album, *Born Gangstaz* (1993). Boss's newfound imagery is seen as a sort of conversion experience in which her middle-class upbringing is rejected for the more visually alluring experiences she describes in detail, such as sleeping on park benches, slinging dope, and hanging out with gangbangers in the process of trying to get a record deal. Through these personal narratives, Boss argues for her placement in the confines of a ghettoized world we have come to regard as "real."

This placement defines the necessary qualifications to participate in a genre in which poverty and violence go hand in hand. Here, the challenge toward authenticity and spectacle coalesces. This celebrated transition from middle-class princess to lower-class female gangsta is best emphasized through a line from her first single, "Deeper" (1993), in which she details her journey "tryin' to get to Watts, but I'm stuck in Baldwin Hills." Watts, the ghetto enclave that formerly defined a large segment of lower-class Black Los Angeles existence, which has recently become populated by as many equally distressed Latinos, is seen as preferable to the bourgeois trappings of Baldwin Hills, the area often racially demarcated as the "Black Beverly Hills." Boss's reversal of desired class affiliation is further endorsement for the superiority of image in understanding the contemporary phenomenon of gangsta culture in a visually literate postmodern landscape. The image of the gangsta has ultimately taken the place of the "real" life circumstances that define Boss's middle-class childhood.

Gangsta Rap and the Hyperreal

This fascination with the image has become one of the thematic calling cards for this particular genre. In a way much like Jean Baudrillard's (1983) description of the hyperreal, a situation in which the "contradiction between the real and the imaginary is effaced" (p. 142), gangsta rap continually blurs the line between fiction and nonfiction. News headlines about the alleged criminal activities of rappers Dr. Dre, Tupac Shakur, and Snoop Doggy Dogg have received a great deal of attention. At one particular time, Dr. Dre was in legal limbo with five court dates pending on as many different charges, all having to do with some violent behavior, including his attack on female rap talk show host Dee Barnes. He is currently under house arrest for violation of his parole. Tupac Shakur has also been charged with several crimes and was convicted for the sexual assault of a New York woman in 1994. During the trial for this most recent conviction, he was shot repeatedly while on his way to record some tracks in a New York studio.

Most notable, however, is Snoop's recent trial and acquittal on murder charges. Snoop, like Tupac, coexists alongside a musical genre that often celebrates a gun culture gone crazy. In a sense, Snoop's track "Murder Was the Case" (1994) sounds like a firsthand account of his real-life exploits. His performance of this tune on the 1994 MTV video awards concluded with a bold declaration of "I'm innocent, I'm innocent," bringing even more confusion about whether he was talking about his own murder case or the song itself—the fiction of the song or the reality of his life.

Important here for the sake of analysis in addition to the actual charges of crime are the solidification of an image and a subsequent celebration of that image, bolstered by any demonstration of this activity in real life. The hyperreal creates a media image that directs attention away from the actual occurrences and thus puts us in a realm of pure spectacle.

To focus on the "real" alone misses the point. People are encouraged to read beyond the literal. The focus must be on the hyperreal or, as Guy DeBord (1983) says, in contemporary society "everything that was directly lived has moved away into a representation" (p. 1). In other words, there is no "real" other than that represented through imagery. In the case of this collusion between image and reality in gangsta rap, we must begin to approximate the significance of what is configured as the ultramasculine identity of Dre, Tupac, or Snoop through representation as opposed to being locked into a fruitless argument as to the relative morality inherent in social actions. Otherwise, we simply become moral arbitrators, rather than critics who can acknowledge these moral concerns but who do so in a way that is also empowered with the ability to explicate the most subtle nuances of cultural production.

The multimediated context in which gangsta rap exists, combined with the discourse of the hyperreal, circulates throughout much of American popular culture and makes this investigation useful. As an extended example of how this functions in relation to larger cultural issues, one need only notice the popularity of daytime talk shows and their reliance on personal narratives that constantly veer between the grotesque and the perverse. The more sensational the story, the more likely that it will be presented as a type of televisual autobiography.

In prime-time broadcasting, viewers constantly see the news magazine format as a vehicle for a certain truthfulness in light of the otherwise fictional nature of most prime-time programming. These news magazines are extensions of this hyperreal phenomenon. The supposedly nonfiction content of these programs separates them from most other forms of prime-time network television. But as the highly publicized situation between NBC and General Motors over the network's rigging an explosive to visually substantiate its point on exploding vehicles demonstrates, we can never be too sure whether we are experiencing the real or the imaginary. Thus, we are in a constant state of the hyperreal once again.

This circulation of the hyperreal aesthetic operates both in mainstream society and in the cloistered world of gangsta culture. As author Brian Cross (1993) explains, "for new generations the culture produced by the seeming hopelessness of the gangsta lifestyle has become the social realism of the nineties" (p. 32). With one of the most popular images of Black masculinity being the gangsta, it is necessary to unravel this complex scenario to understand the simultaneous infatuation and repulsion that has characterized the public's response to this cultural movement.

Way Out West

As rap music gained public momentum throughout the late 1980s, it was still a primarily East Coast movement, with various areas of New York City serving as the breeding grounds for this expanding genre. Yet as the music began to evolve from an often redundant vehicle for male posturing into a discursive arena that foregrounded political concerns, it also began to slowly move away from its New York base into other areas of the country, most notably, Los Angeles and the West Coast.[3]

In the early 1950s, a number of White jazz musicians emerged on the West Coast, playing music that was radically different from what had become the rage of New York—bebop. The proliferation of the "cool" sound on the West Coast, borrowed from the innovations of Miles Davis's album *The Birth of the Cool* (1949, released in 1953), was a rejection of East Coast jazz as a truly Black idiom of expression. Instead, West Coast jazz attempted to create a space for White articulation without the burden of having to deal with the overt "Blackness" of what had become jazz's dominant discourse. The geographical distinction between New York, especially Harlem, and Los Angeles is instructive in understanding the vast difference between the music being played on either coast. The West Coast was indeed thought of and perpetrated as a "White" space of opposition.

The move of rap westward in the late 1980s is similar to this jazz history at the level of a bicoastal transition. As the specter of the West Coast came to prominence throughout the 1980s as a postmodern alternative to the historical dimensions informing the East Coast, especially in technology and culture, it was inevitable that the emergence of a distinctly West Coast African American voice would emerge in the process. As the implied distance between New York and Los Angeles indicated in jazz, the musing of the West Coast would be quite different from what had become common on the East Coast. As a matter of fact, one of the early underground hits of the gangsta rap movement, "Boyz n the Hood" (1987), from which John Singleton drew the name for his first film, was originally written to be performed by the East Coast rap group HBO. According to Dr. Dre, the song's producer, HBO rejected the tune on the basis of its being "some west coast shit" (quoted in Cross, 1993, p. 187).

Unlike the shift westward in jazz, which indicated a racial dichotomy between Black and White styles and musicians, the split between East and West in rap signified an internal dynamic that took place solely within African American culture. With an extension of this argument regarding gang culture itself, Monster Kody (Shakur, 1993b) reveals that "today, Crips are the number one killers of Crips" (p. 19). This internal war of the Crips, or "set trippin'," as it is referred to, suggests the impact of violence among

people who are united by gang affiliation, who supposedly have something in common—
thus, the popular media phrase "Black-on-Black" crime. In the same way that set trippin'
locates conflict internally, we can see the dynamic of differences between East and West
Coast notions of Blackness.

This internal split that defines Blackness through rap music is clearly exemplified in
East Coast rapper Tim Dog's assertive title "Fuck Compton" (1991) and Rodney-O and
Joe Cooley's definitively titled response, "Fuck New York" (1993). In both cases, the
identification of urban space is instrumental in defining identity. The specifically internal
definition serves as a rejection of the often unifying metaphor of race in this context.

It is now apparent that this shift to the West Coast in rap allowed for the articulation of
a particular voice that used a geographical specificity that was considered unique and
"authentic" to one's experiences in the gang-dominated culture of South Central Los
Angeles. The move to the West Coast and subsequent emergence of gangsta rap can, at
one level, be read as a call to specificity that openly critiques the way in which race is
often treated as a monolith in mainstream media representation.[4]

This music was an assertion of one's individuality—and what better way to articulate
individual experiences than to be specific as to the exploits of one's immediate environment
or hood? What could be more individualized and, in turn, argued for as authentic than the
personal experiences of one's own communal existence? According to Tricia Rose (1994),
"identity in hip-hop communities is deeply rooted in the specific, the local experience,
and one's attachment to and status in a local group or alternative family" (p. 34).

Thus, the specific neighborhood or set, often encompassing a geographical boundary
as small as an individual street—for example, the Eight Tray Gangsta Crips represent 83rd
street between Florence and Normandie—became the locus of identity that distinguished
various factions within gangsta culture. The crossing of boundaries, both real and imaginary,
became a primary way of understanding the importance of space and signified a politics of
location that makes gangsta culture another in a long line of African American cultural
products that see the acquisition of space as fundamental to identity formation.

Public Enemy's release *It Takes a Nation of Millions to Hold Us Back* (1988), a record
often thought of as the canonical text for political rap, was gaining mainstream popularity,
whereas in the same year N.W.A.'s second release, *Straight Outta Compton,* was targeted
by the FBI as encouraging the murder of police. Although Public Enemy's album is about
the popularizing of political culture, N.W.A.'s is about the state acknowledging what they
see as the threatening iconography of gangsta culture. As rap was establishing itself as a
contemporary vehicle for political expression, it was also showing early signs of the move
away from this East Coast-based agenda and at the same time solidifying its position as
a coming cultural attraction that would eventually replace this popular political dimension. Letters of reprimand from the FBI worked to strengthen a notorious identity as
opposed to removing this form of identity from the cultural landscape.

Like the quintessential "bad nigga" of African American folklore, the gangsta rapper
grew to be simultaneously celebrated and feared, both in the outside world and by those
in the community. Ice Cube boldly proclaims himself "the bitch killer/cop killer" in "The

Nigga You Love to Hate" (1990); it is obvious that he sees the outside oppression of the police as equally repulsive as the presence of women, especially Black women. The gangsta rappers, like the bad nigga, find community only among themselves, and even that bond is constantly being challenged.

Thus, in 1988, as the political edge of rap was taking hold of popular imagination, opening up discussions surrounding a resurgent Black Nationalism and ideas of Afrocentricity, gangsta rap was slowly emerging as a defiant form that not only would threaten those on the outside, here represented by an overzealous FBI, but also would eventually become the total antithesis to any sustained political discourse in popular culture. In this sense, gangsta rap is, in many ways, about the death of political discourse in African American popular culture.

The Revolutionary Lumpenproletariat

Yet an inevitable link between politics and "gangsterism" cannot be avoided. As Mike Davis (1990) has argued in his provocative *City of Quartz,* the original Crips were an offshoot of the Black Panther Party. Davis, who consistently references Donald Bakeer, the author of *South Central* (1987), the book from which the Steve Anderson film by the same name was adapted, states that the original acronym CRIP stood for "Community Revolution in Progress." Davis goes on to say that the Crips "inherited the Panther aura of fearlessness and transmitted the ideology of armed vanguardism (short of its program)" (p. 299).

Throughout her autobiography and revisionist history of the Panthers, *A Taste of Power,* Elaine Brown (1992) suggests that many members of the Party were themselves gangsters at one time or another. This, coupled with the legend of Malcolm X's life in the gangster underworld prior to joining the Nation of Islam, gives strong indication that there has always been a historical link between the subversive tendencies of Black gangsters and their easy transition to a position in the political vanguard. This tendency represents what Davis has called the *revolutionary lumpenproletariat.*

Most representative of this transition from gangster to political ideologue is West Coast rapper Ice Cube. As mentioned earlier, however, not only does Ice Cube inform the genre of gangsta rap, but also his image, especially his politics, in an otherwise apolitical environment, functions as a metacritique as well. Originally a founding member of N.W.A., Ice Cube was the primary writer for most of the group's early material, most notably, "Fuck tha Police" (1988). Ice Cube's departure from the group in 1989 has become an often-discussed highlight in the history of rap music. If N.W.A.'s *Straight Outta Compton* is the record that brought nationwide attention to the developing genre of gangsta rap, then Ice Cube's first solo effort, *Amerikkka's Most Wanted* (1990), demonstrated that this music was going to be more than a passing trend. The release of the album also began the cultural process that would make Ice Cube one of the few rap superstars.

Most intriguing about the musical life of Ice Cube is the evolution that he has undertaken from the days as the lyricist for N.W.A. to his most prominent position only a few short years later. Ice Cube's popularity and longevity are quite important, considering that many rap acts have been popular on a single or maybe an album but that few have a body of work that includes five albums; the extended play record *Kill at Will* (1991); lead production credits on three YoYo records, *Make Room for the Motherlode* (1991), *Black Pearl* (1992), and *You Better Ask Somebody* (1993); Mack 10's debut release, *Foe Life* (1995); and an extremely popular Da Lench Mob album *Guerrilla's in the Mist* (1992)—not to mention his early work with N.W.A.

In a music industry replete with one-hit wonders, Ice Cube's continued success is quite relevant in understanding the transition of gangsta rap from marginal to mainstream—a transition he had a great deal to do with. Ice Cube even alludes to this on his 1992 album *The Predator* when asking the self-reflexively rhetorical question, "nigga wit the third album, how come you don't fall off?"

The transition from pure gangster to ideological gangster is not a simple one. His first solo record, *Amerikkka's Most Wanted,* demonstrates the unabashed narrative qualities that distinguish Ice Cube from all competitors. This album is a first-person account of the trials and tribulations involved with living in South Central Los Angeles. His storytelling ability is truly in the same league as a Richard Wright or Ralph Ellison, adapted to fit the modern medium of rap music and contemporary society. With as much detail as one can convey lyrically, Ice Cube forces his familiarity of the hood on the listener in such an affective way that various areas of South Central become as common a public Southern California landmark as Rodeo Drive or the ubiquitous Hollywood sign.

The album opens with Ice Cube being marched to the electric chair for execution. When asked if he has any last words, he responds, "Yeah, I got some last words, fuck all y'all," at which point listeners hear a loud "switch" indicating that he has now been executed. Immediately following this execution, he goes on to castigate multiple areas of American society in the provocatively titled "The Nigga You Love to Hate," ultimately leaving no stone unturned. His targets included, among others, sell-out rappers, mainstream African American celebrities, police, women in general, and anyone else who happens to get in his way.

For instance, one of the most notable lines on the record contains a quick deconstruction of the escalating fame of late-night talk show host Arsenio Hall, "they ask me if I liked Arsenio/bout as much as the bicentennial." His open contempt for women is noticeable as he increases the intensity of this opening salvo. His all-encompassing rejection of mainstream society parallels many of the opening solos rendered by John Coltrane during the latter part of his musical life. Ice Cube has repeated this opening strategy on all his ensuing projects.

Although this first album clearly set in place Ice Cube's utter defiance of White racism and African American complicity, it would be the only album of its type from Ice Cube. *Amerikkka's Most Wanted* represents the perfect distillation of gangsta culture: a nihilistic celebration of the aggressive excesses of Black male ghetto life presented in its most sensational fashion without apologies or any misgivings. Although this record, like *Straight Outta Compton* before it, is a detailed cruise through the mind of a young "G," the representation would be short-lived. His next album, *Death Certificate* (1991), clearly took a much more politicized stance while remaining true to its gangsta rap roots.

To fully appreciate the critical possibilities of *Death Certificate,* one must begin with the album cover, which shows an enthusiastic Ice Cube with his right hand firmly placed over his heart in mock reverence, standing over a dead body lying on a gurney. The body is wrapped in the American flag, and the toe tag identifies the body as that of "Uncle Sam." The metaphoric death of "America" and all it stands for clearly indicate that a more political direction is intended on this record. Ice Cube furthers this specific theme in the tune "I Wanna Kill Sam," in which he declares, "I wanna kill Sam/cause he ain't my muthafucking uncle." *Death Certificate* suggests Ice Cube's coming to consciousness in the gangster-to-revolutionary tradition of both Malcolm X and the Black Panther Party.

In addition to being a concept album, akin to Marvin Gaye's *What's Going On* (1971), the record is divided into the "death" and the "life" sides. Ice Cube begins the album with this proclamation: "Niggaz are a state of emergency/the death side/a mirrored image of what we are today/the life side/a vision of where we need to go/so sign your death certificate."

The death side is a continuation of *Amerikkka's Most Wanted* in that it chronicles the devastation and social ills that have captivated the ghetto landscape of South Central Los Angeles. The topics range from overt sexism in a story about sexually promiscuous women ("Giving up the Nappy Dugout"), the rampancy of sexually transmitted disease ("Look Whose Burning"), the constant need to be armed or "strapped" in the volatile ghetto ("Man's Best Friend"), and the exportation of gang and drug culture to the Midwest and other areas of the country ("My Summer Vacation"). Yet the life side takes this album to another level. The life side is a series of societal critiques that deal with, among other things, White male stereotyping and the sexual harassment of Black women ("Horny Lil Devil"), selling out in general ("True to the Game"), the constant danger of gang life ("Color Blind"), and a self-reflexive critique of Black people's own participation in their oppression ("Us"). This side also contains a controversial track titled "Black Korea," which anticipates the African American versus Korean American conflicts that began to receive widespread media attention in the aftermath of the L.A. uprising in April 1992.

Most important about this side is the influence the teachings of the Nation of Islam have had on Ice Cube's thinking. With several references to the language and icons of the Nation, Ice Cube's political position stands out. It is clear that he has become an advocate of a historically ideological position, but he nuances this ideology in a fashion specific to the genre within which he exists. As an overt indication of this fusion of gangsta iconography and Nation of Islam theology, one need only look at the photo insert contained inside the compact disc packaging.

First of all, there is a noticeable difference in his appearance from the first album, whose cover had an Ice Cube dressed in all black, wearing a skull cap with the ends of his Jheri curl peeping out from the rear. He is standing in front of several similarly clad members of his crew in L.A.'s historic garment district. On the cover of his second album, Ice Cube has shaved his head, having cutting off his infamous Jheri curl, a hairstyle that had long faded from popularity everywhere but in South Central Los Angeles. This open defiance of accepted style had been his trademark on the first album and the trademark of many L.A. gang members as well. Yet on *Death Certificate,* he openly rejects this style all together—"cut that jheri juice/and get a bald head/then let it nap up." Ice Cube's rejection of a chemically treated hairstyle, similar to Malcolm X's rejection of his "konk," is no longer considered consistent with Cube's changing image and his coming to Black consciousness, whereas the bald head of an African American man, from boxer Jack Johnson forward, has always connoted an empowered virile image of masculinity. The bald head had, at the time of this album, once again become a popular hairstyle among many African American men, especially athletes and rappers. His reference to letting it "nap up" further signifies his recognition of a Black aesthetic and refusal to accept dominant standards of beauty.

In addition to this new appearance, the cover of *Death Certificate* shows Ice Cube reading from *The Final Call,* the Nation of Islam's biweekly newspaper. He is centered in the middle of the picture; the image is divided into two groups on either side of him. On one side are several members of Ice Cube's "posse" Da Lench Mob, dressed in various phases of "G" style. On the other side are several suit-and-bow-tie-adorned members of the Fruit of Islam, the security force for the Nation. The headline on *The Final Call* reads "UNITE or PERISH," and Ice Cube's presence symbolically attempts to unify these two separate groups.

This visual unification is what underscores the thematic unity of the album itself. The coupling of gangsta iconography and the nation's own specific brand of Black nationalism would eventually define Ice Cube's image. Like members of the Black Panther Party, Ice Cube was a gangster who had come to a specific political consciousness from his location on the ghetto streets. Yet in the same way as the Panthers, especially in light of Elaine Brown's (1992) revisionism, he nonetheless remains a part of his original gangsta community. Locating himself in one camp or the other would not be that illuminating, but an oscillation between the two poles of identity distinguishes Ice Cube as both a rapper and a popular icon of contemporary Black male existence. It is clearly impossible to deny Ice Cube's impact on the genre, but it is also important to point out that his significance in crossing the boundaries between gangsta and ideologue is not shared by other rappers.

What has happened with L.A. gang culture, after the FBI and the L.A. Police Department's concerted effort to annihilate any form of empowered political articulation from the community, is a reversal of the previous transition from gangster to revolutionary. Instead, the onetime revolutionaries accepted the violent tropes of political resistance without the informed ideology that separates gangsters from freedom fighters. In a similar fashion, the move from the politicized domain of East Coast rap championed by Public Enemy to the celebratory nihilism of the West Coast gangsta rap demonstrates the current death of acknowledged political activism in contemporary culture.

Ain't Nothing But a "G" Thang

This political decline is also quite consistent with the position advocated by Dr. Dre. In an interview with Brian Cross (1993) concerning his initial thoughts regarding his entrance into gangsta rap as a producer, Dre says, "I wanted to go all the way left, everybody trying to do this black power and shit, I was like let's give 'em an alternative" (p. 197). The alternative comes in the form of celebrating the antithesis of politics through gangsta rap. He goes on to say, in reference to the purpose and function of making records, "it ain't about who's the hardest, it's about who makes the best record, as a matter of fact it ain't even about that, it's about who sells the most records" (p. 197).

These open statements, which denounce both politics and the celebration of craft for the bottom line of making money, are creatively put into his music in "Dre Day" (1992), in which Dre emphatically announces "no medallions/dreadlocks/or Black fist/It's just that gangsta glare/with gangsta rap/that gangsta shit/brings a gang of snaps." These obvious symbols of Black nationalism are completely rejected in favor of celebrating capital (snaps), excess, grotesque spectacle, and the ultimate nihilism that pervade gangsta culture. Dr. Dre's attitude toward the function of rap in the music business and his attitude toward Black nationalism in general clearly state the politics of an antipolitical movement.

This eager embrace of the excesses of capitalism places Dre and his contemporaries right in the midst of arch conservative proponents of an unadulterated free market economy. This argument is supported by rapper Easy E (who later died of AIDS) and his much discussed appearance at a Republican party fund-raiser in 1991. Yet as Easy himself asked rhetorically, how can someone make a song such as "Fuck tha Police" and be a Republican? This reductionistic reading is ultimately too simplistic to accommodate the contradictions of this approach. Another view could also be used to locate Dre's position, like that of the underworld gangsta, on the axis of a capitalist metacritique, which sees the manipulation of capitalism by African Americans, both in art and business practices, as a perverse deconstruction of an otherwise exclusionary monopoly enjoyed primarily by White male corporate interests.

Dre's rise to prominence in the world of gangsta rap highlights the salient qualities that ultimately define this most effective cultural movement. Thematically, Dre's style features the glamorization of wanton violence, a strong propensity toward self-indulgence, and the exaggeration of these forms through the use of spectacle. This reliance on spectacle is most demonstrative when studying the overall production values of the work, especially as it relates to an overt theatricality. *The Chronic* (1992) features several intermittent pieces of comedic dialogue, many of which openly reinscribe the marginal position of women in this world, as well as re-creating many episodes of popular culture, although specified to the gangsta idiom. A good example of this is the "$20 Sack Pyramid," in which a group of contestants, in the style of a popular game show, compete for a bag of marijuana and a trip to the Compton swap meet. This type of comedic skit is combined with the recurrent use of special sound effects and the overall themes of violence and otherwise debased pathologies. It also uses a nuanced reading of the music of George Clinton, which works to define Dr. Dre's popularity. He has tapped into several illuminating areas of popular culture in a way quite similar to voyeuristic pleasure associated with the films of Quentin Tarantino, which have become a recurring self-reflexive cycle of pop trivia taken to its most extreme form. Once again, excess and the pushing of the societal boundaries of acceptable taste have come to augment a large sector of what now exists throughout the cultural landscape.

Dre has taken the production end of rap into an entirely different arena than had previously existed, for once making production somewhat superior to the lyrical styling of the rappers themselves. In a medium that has always foregrounded the rapper, Dre has made the producer a significant part of the overall equation. This situation is itself enhanced by Dre's partner Snoop Doggy Dogg, whose raps provide the oral legitimacy in an unquestioned fashion. The production values for both Dr. Dre's record *The Chronic* (1992) and Snoop's *Doggystyle* (1993) are quite similar, firmly rooted in Dre's landmark "G funk" style, a style that uses the imagery of the "G" in conjunction with funk-inspired beats of the 1970s, especially those of George Clinton. This combination of popular imagery and familiar grooves from a previous era assist in making the G funk culture so popular on such a mass scale.

Both Dre and Snoop realize the importance of video in promoting records outside the community and in the postmodern age. Each of their music videos provides a slice of life relative to the day-to-day experiences of lower-class Black life in South Central and Long Beach. As MTV began listing the directors' names during the playing of music videos, this recent form of popular culture took on cinematic overtones, foregrounding the visual side of the equation and allowing individuals such as Dr. Dre the opportunity to create their own visual images along with the illuminating lyrics that had been the defining component of the music.

Thus, a video such as the popular *Ain't Nothing But a "G" Thang* (1992) demonstrates Dre's directorial skills along with his noted musical production. Dre has used the video format as a central means of expression that can articulate the nuances of ghetto life in a seemingly authentic fashion. For instance, *G Thang* functions much like an ethnographic excursion into the hood by highlighting such things as a barren front lawn, a gun tucked neatly in the small of someone's back, a refrigerator full of Olde English Malt Liquor, and a festive community picnic during which Snoop and Dre ride through South Central in their low riders. In a similar fashion, *Let Me Ride* (1992) features various local landmarks, such as a popular hand car wash frequented by several "ballers," along the Crenshaw District, as well as other areas throughout South Central.

The localizing affect allows Dre to clearly link himself with the community that is vital to his musical authenticity. This link also provides a specialized view of Los Angeles quite different from other forms of popular representation. The ability to explicate the nuances of a local community and forge this explication with musical distinction provides for something akin to an ethnographer's oral history, yet in this case the ethnographer is part of the community being represented. Dre's use of the video medium and his subsequent direction of the short film *Murder Was the Case* (1994) demonstrate the way in which gangsta rap has merged with the visual means of production to provide an alternate voice in light of the film industry's overall hesitancy regarding a full-fledged development of Black cinema. With the music as the foundation for this excursion into pop culture at a larger level, music has once again been the motivating factor behind advances in other areas of culture. That these new ventures are specific to a racialized reading of Los Angeles in the postindustrial era makes the overall situation that much more illuminating in regard to the representation of race, class, and Black masculinity in contemporary urban America.

Conclusion

Gangsta rap gives a vivid picture of the inner workings of identity politics specific to Black masculinity within the contested cultural space of Los Angeles. Although the tendency has been to criticize this music for its celebration of violence, misogyny, and an overall nihilistic approach to life, I have suggested that the relevant issues also include understanding the cultural currents that shape the thematic content.

Gangsta culture mirrors other forms while still distinguishing itself. It is clearly influenced by the most extreme spectacle of Hollywood; here, the metaphoric closeness between the industry and the various hoods of Los Angeles cannot be denied. Gangsta culture is clearly informed by the visual, yet the actual images that circulate through the lyrics of the music lend themselves easily to the production of visual culture, as Dr. Dre's direction of *Murder Was the Case,* which is based on a Snoop Doggy Dogg song, indicates.

In the same way, Dr. Dre and Ice Cube's collaboration *Natural Born Killers* (1994) takes its cue from Oliver Stone's film by the same name yet puts their own spin on the culture's overall obsession with violence. The irony of the whole thing is that Stone's film is seen as a critical meditation on violence, whereas the collaboration is thought of as encouraging this pathology. The perverted racial reasoning that informs this perception becomes less significant when the full spectrum of gangsta culture, its exposition on lower-class Black masculinity, and its placement in contemporary Los Angeles is brought to attention. Thus, the full exposition of the culture becomes paramount to understanding the relevant issues within a form that is either neglected or misread by a rather limiting moral critique.

Notes

1. Another exploration of this subject exists in Robin Kelly's (1995) chapter "Kickin Reality, Kickin Ballistics: Gangsta Rap and Postindustrial Los Angeles" in *Race Rebels.* Although Kelly covers some of the same terrain, his argument attempts to resurrect a political image of gangsta rap while seemingly denouncing as regressive most of the material that does not fit into his discussion. A rejection of Snoop Doggy Dogg's presence at the conclusion of his chapter is indicative of this strategy. It is my argument that the great majority of gangsta rap is defined by the rejection of politics that Kelly castigates; thus, I am more interested in these overt rejections that evolve in the genre and eventually make the genre the major market commodity that it is, especially post-Ice Cube.

2. For a lengthy discussion of the "bad nigga," see Lawrence Levine's (1977) *Black Culture and Black Consciousness* and Charles Henry's (1990) *Culture and African American Politics.* For a discussion of the reoccurrence of this image in contemporary popular culture, see "The Day the Niggaz Took Over" (Boyd, in press).

3. Although Tricia Rose's (1994) *Black Noise* is an important contribution to the discussion of rap music, one of the central problems with the book, for me, is its apparently chauvinistic view of New York as the exclusive center of rap as a discursive form. Although this was true in the 1980s, West Coast rap has transformed the medium since that time both in cultural impact and in record sales. Gangsta rap is what made rap into the mainstream force it is today. Thus, my focus here is on the specificity of West Coast rap in creating its own discourse and ultimately influencing and forcing the East Coast to reconsider its onetime dominance in the rap industry. (See also Boyd, 1994a.)

4. For a further discussion of the differences between East Coast and West Coast style in rap music, see "The Geography of Style" (Boyd, 1994b).

References

Bakeer, D. (1987). *South Central.* Inglewood, CA: Precocious.

Baudrillard, J. (1983). *Simulations.* New York: Columbia University Press.

Boyd, T. (1994a, Autumn). Check yo self, before you wreck yo self: Variations on a political theme in rap music and popular culture. *Public Culture, 7*(1), 289-312.

Boyd, T. (1994b, October 2). The geography of style: Rap wages a bicoastal battle over the nation's hearts, minds, and ears. *Chicago Tribune,* pp. 33-34.

Boyd, T. (in press). The day the niggaz took over: Basketball, commodity culture, and Black masculinity. In A. Baker & T. Boyd (Eds.), *Out of bounds: Sports, media, and the politics of identity.* Bloomington: Indiana University Press.

Brown, E. (1992). *A taste of power: A Black woman's story.* New York: Pantheon.

Cross, B. (1993). *It's not about a salary: Rap, race, and resistance in Los Angeles.* New York: Verso.

Davis, M. (1990). *City of quartz: Excavating the future in Los Angeles.* New York: Verso.

DeBord, G. (1983). *The society of the spectacle.* Detroit: Black and Red Press.

George, L. (1992). *No crystal stair: African Americans in the city of angels.* New York: Anchor.

Henry, C. (1990). *Culture and African American politics.* Bloomington: Indiana University Press.

Kelly, R. D. G. (1995). Kickin reality, kickin ballistics: Gangsta rap and postindustrial Los Angeles. In R. D. G. Kelly (Eds.), *Race rebels.* New York: Free Press.

Levine, L. (1977). *Black culture and Black consciousness.* New York: Oxford University Press.

Pulley, B. (1994, February 3). How a nice girl evolved into Boss, the gangster rapper. *Wall Street Journal,* pp. A1-2, A10-11.

Rose, T. (1994). *Black noise: Rap music and Black culture in contemporary America.* Hanover, NH: University Press of New England.

Shakur, S. [Monster Kody Scott]. (1993a, April). Can't stop, won't stop: The education of a Crip warlord. *Esquire, 119,* 87-139.

Shakur, S. [Monster Kody Scott]. (1993b). *Monster: The autobiography of an L.A. gang member.* New York: Grove/Atlantic.

Wallace, A. (1993, April 4). Making monster. *Los Angeles Times Magazine,* 16-56.

West, C. (1993). *Race matters.* Boston: Beacon.

Milagros

RUBÉN MARTÍNEZ
(para Rubén Ortíz)

*(con el perdon de la virgencita que tanto adoro y que seguramente entiende
el lenguaje metafórica del texto que sigue . . .)*

Milagro en México
and miracle on Broadway
he aquí la Milagrosa
behold the Miracle worker:
Nuestra Señora la Reina de Los Ángeles de Porciúncula

She is la abuela hawking mangos singing her L.A. barrio mambo
    habiendo llegado soñando Hollywood
She is the Indian boy from Oaxaca with paint-flecked jeans
    thumbing for work on Santa Monica Boulevard
She is the single mom from Tijuana whose fingers bleed
    onto the white cloth stitched by the Singer Electric
She is the rockero from la capirucha now in East Los temblando
    con la fiebre del amor in los tiempos del AIDS

La Milagrosa
we cling to your glowing robes
admiramos tus labios tibios
your olive skin tu peluca rubia
(puede ser el trasvesti mamando la verga del gringo
    que paga buena lana)

She is our salvation!
Our Chicano-Salvatrucho-raza-pusmoderna self-realization
on loan from Tenochtitlán
thriving on the jarocho raps of the jornaleros
¡wacha el LAPD!
She is bought and sold and indentured
She is recognized as an ally of Malcolm X and the Chicano troops
    at Guadalcanal
She was seen saqueando las tiendas con la turba llevándose
    montones de Pampers y gritando consignas in support of
    Rodney King
No Justice No Peace!
She graces the hoods of lowriders is tattooed on the breasts of
    quinceañeras and She lays bleeding after the drive-by shots
    ring out

Como el caballo blanco of corrido fame
She came North but She was already here
on our 1848 streets, on La Ciénega Boulevard, on La Brea
    Boulevard, on the Cahuenga Pass . . .
and She has turned Her loving gaze South time and again,
cargando al Elvis Presley hasta Insurgentes
y meneando Sus caderas al compás del rap en Spanglish
She stars in all the films that make
the machos and the kids the abuelas and the cojos
yearn to brave la migra and cross la línea cruda
    para llegar al otro lada

She is the Other Side here
and el Otro Lado there
She's la Chicana that gives Chapultepec una migraña
es la guerrillera burlándose del gringo
es el milagro de nuestra esquizofrenia
She's the miracle of our survival

She is more powerful than the PRI; she gets more respect than the narcolords of
the north; she's better loved than even mom or grandmom and her home-cooked rice
and beans; she's at once a feminist and a bastion of family values; she's an ethereal
being who's as friendly as your aunt, and she's the only national figure capable of
turning out crowds big enough to shame any "politician of the people."

She is the *Virgen* de Guadalupe, the "Mother of Mexico," which means that without
her Mexico wouldn't exist. And, judging by the fervor with which the faithful turn out
to venerate her, it is to the *Virgen*—and not the politicians—that Mexicans are turning
for hope in this turbulent time.

There is precious little to be certain about in Mexico these days. There's turmoil in
Chiapas, and political assassinations have left the citizenry shaken and cynical. NAFTA
has yet to deliver economic miracles, and the passage of Proposition 187 in California
gnaws away at the hopes of millions of Mexicans who sought a way out of crushing
poverty by working in the north.

There is precious little to be certain about in Mexico today, except the *Virgen* de
Guadalupe.

I join the pilgrims that come from all over the republic to celebrate her feast day,
December 12, but the party begins the night before. A few arrive by plane, but the
vast majority arrive by bus, car or motorcycle, and, especially, bicycle. Yet others come
jogging from distances greater than the Greek marathon, and walking 10 or 100 miles,
carrying banners and picture frames and ornate wooden altars, all with her image.

They come as they've come for a thousand years, to the Cerro del Tepeyac, the
modest hill to the north of Mexico City that was once home to the Aztec goddess
Tonantzin, the serpent-woman, the mother of All Life, and, which since 1531, has been
home to the Catholic *Virgen* de Guadalupe. It was here that she appeared to the Indian

neophyte Juan Diego; here that she told him to ask the Spanish Catholic authorities to build a chapel to venerate her. It is no coincidence that the Guadalupe appeared on Tonantzin's old turf. For the Guadalupe is the Indian *Virgen:* the spiritual marriage of Indian polytheism and European Catholicism. Without her, Mexico would not have survived the Conquest. The *Virgen* is the essence of Mexico—a constant cultural tension and cooperation between its Iberian and Indian selves.

The pilgrims come, and they come young, a brown nation of Indian and mixed-race heritage—with their hair neo-hippie long or punk short—*mestizo* city kids and *indígenas* from the provinces. Now and again there's the surreal sight of a white European-Mexican wearing an Aztec headdress and rattlers at the ankles, but the more common pilgrim is the Indian-looking rocker wearing a leather jacket and carrying a ghetto blaster bearing a sticker with the likeness of the *Virgen* right alongside Metallica and Nirvana.

They gather in the huge plaza at the foot of Tepeyac, crowned by a 17th-century chapel, next to the abstract architecture of the modern cathedral. The biking pilgrims park tens of thousands of machines on top of one another, ceremonial piles of aluminum and rubber. The pilgrims drink. The pilgrims smoke—both tobacco and that other thing that both Clinton and Gingrich smoked in grad school. The *Virgen* is no stodgy old maid.

It is the biggest party I've ever been to—nearly 4 million pilgrims arrive between Saturday night and Monday morning. My street paranoia fades as I realize that this is probably the safest place on the planet to be: Who would dare dis 'Lupe by committing the sin of robbery? There is virtually no security, except for a few Red Cross crews with stretchers and ambulances at the ready for those who pass out in the crush of the crowd. The few cops are in an uncharacteristically jovial mood. They while away the hours by taking pictures of themselves at Technicolor-bright Polaroid altars depicting the *Virgen*'s apparition and munching on traditional egg and nut breads along with everyone else.

After two in the morning, the night grows chilly and the party-energy flags. The thousands of delegations of faithful turn in for the night—the marathon runners from Hidalgo, the Marian devotees from Morelos, the Indian dancers from Oaxaca, the young former drug addicts-turned-missionaries from Tijuana. They lay out their blankets and huddle together. It's as if the country is in one huge embrace; the country hugs itself through the night.

This is the greatness of *mestizo* culture: Everyone's welcome here—German tourists and Chicano anthropology students, yuppies and street toughs.

We can, in Rodney King's words, all get along.

Through Her.

Because she is Indian and Spanish—Tonantzin and the *Virgen.* She's a rocker and an Aztec dancer. Her skin is olive, smack in between indigenous copper-brown and Iberian white. She's the woman who puts the Mexican macho in his place—no matter how much he beats his chest, She's the origin of All Things—but pities him and offers him succor just the same. Both Pat Robertson and Hillary Rodham Clinton would get along with her famously, because she loves families, big families that stay together, but lashes out at anyone who'd endanger a child's well-being. She is, after all, the Savior's mother and sees her Son's visage in the face of every Mexican son and daughter. Perhaps that's why the pilgrims are so young. It is young Mexico that stares into a bleak future, a violent world. It is young Mexico that is looking for itself today, in the jungles of Chiapas and along the endless asphalt of Mexico City and in the cold cities of the North. It is young Mexico that so desperately needs to believe.

Skyrockets burst in the predawn sky as I climb the Cerro del Tepeyac. A tearful nostalgia overwhelms me: I am returning to my grandmother's home, the prodigal son born in the North returning to wrap myself in the gentle folds of Tepeyac. It's okay that my Spanish isn't perfect, that I eat hamburgers and love rock 'n' roll. All the *Virgen* asks of me is faith.

And I do believe—more in her than in the politics of the country whose passport I bear. Such a proclamation of faith is probably enough for California's 187 forces to call for my deportation ("He swears allegiance to some Catholic witch!"). So be it. Today, I don't mind being told, as I have been so many times in the last year, to "go back to Mexico." Damn right I've come back. And I'll return to Los Angeles with this faith: that cultural contraries can create new societies that are greater than the sum of their parts (a process that the United States still refuses to believe). And that, as Mexicans always say, "*no hay mal que dure cien años*"—there's no malady that lasts a hundred years.

Beginning at dawn on the 12th, thousands of Indian dancers pound the lavastone of the plaza with their bare feet amid clouds of incense. The beat of the drums melds with the church hymns emanating from the cathedral. Pilgrims painfully approach the cathedral on their knees, rosaries swinging from their hands, sweat streaming down their brows.

We come to pay her tribute and to petition her as well. I speak to a family that's come all the way from East Los Angeles to honor her. "I want to ask her for better luck with work in the North," a teenage son, a long-haired rocker, tells me. "And for her to accompany all Latinos in California now that they've passed 187," adds his older brother.

I enter the cathedral for Mass. Procession after procession of Indians make their way toward the altar where the tunic worn by Juan Diego 463 years ago hangs, framed in gold. Tradition holds that the *Virgen* emblazoned the cloth with her own image—a miracle to convince the cynical bishop who'd waved off Diego's message from the *Virgen* as the ravings of a drunk Indian. In green, red, and gold raiment, she looks at you with love and compassion, her hands clasped over her chest to let you know that she is sending your prayers along to the One above her.

In the end, that's what She's all about: faith, which is another word for hope. I've found in Mexico, through Guadalupe-Tonantzin, what I'd lost in Prop. 187 California, in 3-Strikes-You're-Out-America, in Rodney King Los Angeles.

I have faith that Mexico will survive this turbulent time with its essence intact. That essence is the festival of the *Virgen,* where all Mexico's children come together to admit that the very pain of our history—the Conquest, the lingering racism against the Indian, the uncertainty of free trade, the diaspora and conflict of immigration—is what offers us the path toward redemption. It is a hope that brings with it a tremendous responsibility: to live up to the *Virgen*'s own faith in *us.*

I'm not so sure about my other country. There's no *Virgen* to bring us together, and the secular ideals that are supposed to make America work are crumbling in an orgy of antidemocratic legislation. The Mexican-ness of the *Virgen*'s festival—its collective intimacy, its controlled anarchy, its *mestizo* embrace of otherness—is misunderstood in the United States. To them, Mexico is premodern, pagan, dirty, and dumb.

And yet what Americans misunderstand about Mexicans is precisely what they need the most. More than ever, Americans cling to their individual "space"; their generosity grows fickle. Americans need to embrace themselves. They need to tear down the walls they've erected between themselves and the other, in their neighborhoods, within their families.

As the sun sinks into the coppery hues of the smoggy horizon of the most populous city on earth, the festivities of Mexico's most important holiday draw to a close. The Indian elders leave the cathedral, chanting in the Spanish of the *Virgen* and in the dialects of Tonantzin: "*Adiós madre del Cielo,*" Goodbye, Mother of Heaven. They walk backwards, their eyes never leaving the doorway to the cathedral. Mexico will never turn her back on her faith.

Our L.A.? 9
Korean Americans in Los Angeles After the Civil Unrest

EDWARD J. W. PARK

Introduction: Spectators No More

In the preface to his book *Two Nations: Black and White, Separate, Hostile, Unequal,* Andrew Hacker explains why a book on contemporary race relations written in 1992 focuses almost entirely on African Americans and whites. He argues,

> Two Nations will adhere to its title by giving central attention to black and white Americans, and the reason for this emphasis will be made evident. In many respects, *other groups find themselves as spectators,* while the two prominent players try to work out how or whether they can co-exist with one another. (p. xii, emphasis added)

In the same year the book was published, Los Angeles erupted in flames over the acquittal of four white police officers in the case of Rodney King's beating. After nearly a week (April 29 to May 5, 1992) of intense protest, looting, and burning, the most devastating civil unrest in modern U.S. history resulted in the loss of 52 lives, 16,291 arrests, and nearly a billion dollars of property damage (Oliver, Johnson, & Farrell, 1993, p. 119). Among those affected, Latinos constituted the plurality of those arrested, and Korean

AUTHOR'S NOTE: I would like to thank H. Eric Schockman, Karen Umemoto, and Regina Freer for comments and suggestions on earlier drafts. Michael Dear and Curtiss Rooks also provided words of support and encouragement, and Reiko Furuta helped with the final editing. Final thanks go to John Park for his excellent editorial skills and critical insights.

Americans sustained nearly half of all property damage (Chang, 1994, p. 101). For weeks to come, millions of Americans watched armed Korean American merchants and their Latino employees standing guard against the predominantly Latino and African American looters. In a singular moment, "spectators" in Hacker's narrative emerged as central and defining players in U.S. race relations.

From the New York City Riot of 1863 to the Liberty City Riot of 1980, civil unrests have been a permanent and reoccurring feature of U.S. racial and ethnic history. They have occurred most often during times of rapid social change and have powerfully captured the tensions in existing racial realities. Civil unrests have also marked the most profound turning points in U.S. racial history, signaling new directions in U.S. race relations (Marable, 1991). In addition, civil unrests have forged new social identities for those involved. In the 1960s, "Negroes," "Orientals," and "Mexicans" entered that tumultuous decade and became "Blacks," "Asian Americans," and "Chicanos" (Munoz, 1989; Omi & Winant, 1994, pp. 101-112).

As the last visible signs of the Civil Unrest of 1992 disappear, the task of understanding its causes and grasping its consequences remains. One of the first steps toward that task is to recognize that this event marked the transition of the United States from a biracial to a multiracial society: The spectators have now taken center stage. In addition to the inclusion of Asian Americans and Latinos in the U.S. racial drama, the Civil Unrest of 1992 marks a transition in U.S. racial history in another way. Since the 1960s, when the Watts Rebellion of 1965 focused the nation's attention on the problem of racial inequality, a host of legal, economic, and political changes have profoundly reshaped the social terrain in which race relations and politics unfold (Omi & Winant, 1994, chap. 6). If the driving force of civil unrests of a generation ago was rooted in the simplicity of white oppression of racial minorities in the context of *herrenvolk* democracy, the new driving force behind contemporary race relations is found in the complexity of intergroup and multigroup relations within the context of post-civil rights politics (Lyman, 1974; Omi & Winant, 1994, p. 144). As with previous civil unrests, the full impact of the Civil Unrest of 1992 will be found only after the political changes unleashed from that event unfold and a new political trajectory is established. In Los Angeles today, the intensity of that process is exceeded only by the stakes involved. And, as with other civil unrests, all the major participants will be transformed.

This chapter first seeks to contextualize the Civil Unrest of 1992 by discussing the changes in racial politics of Los Angeles from the 1970s to the 1990s. Although the discussion is largely limited to Los Angeles, many of the theoretical arguments and empirical observations are relevant to other U.S. cities that have undergone similar political and demographic changes. Second, the chapter describes the specific politics of Los Angeles after the Civil Unrest by focusing on two dominant groups—the "Establishment Coalition" and the "Progressive Coalition"—that have emerged in the city's political struggle for power. In their mobilization, they demonstrate the importance of multiracial coalition building in the politics of a "new" Los Angeles. Third, I conclude with a discussion of the L.A. experience to draw out some of its lessons. In all the remaining sections, the role of Korean Americans will be underscored. Through their political involvement, they reflect the depth of political change in Los Angeles from a generation ago, the centrality of multiracial coalition politics since the Civil Unrest, and the prospects for racial politics in the future. In Los Angeles today, Korean Americans are no longer spectators.

The Setting: Demographic Changes, the Eclipse of the Civil Rights Coalition, and the Korean American Community

Any discussion of contemporary politics in Los Angeles must begin with the tremendous demographic changes the city has undergone in the last several decades. From 1970 to 1990, the population of whites in Los Angeles declined from 59.0% to 37.3%, and the African American population declined from 17.7% to 14%. During the same time, the Latino population more than doubled, increasing from 18.3% to 39.9%, whereas the Asian American population increased from less than 5% to 9.8% (Sonenshein, 1993, pp. 87, 247). The most important cause of the demographic shift was the immigration and reform laws of 1965 (Hing, 1993, pp. 18-19), which removed racial bias in U.S. immigration policy in favor of economic and humanitarian concerns, especially family reunification (Reimers, 1985). As the full force of these immigration and reform laws took hold in the 1970s, international immigration from Mexico, Central America, and Asia transformed Los Angeles, the destination point for the largest number of post-1965 immigrants (Barringer, Gardner, & Levin, 1993, Chap. 4). By 1990, Latinos surpassed Anglos as the largest racial group in the city, and the number of Asian Americans in Los Angeles County surpassed that of African Americans (Ong & Azores, in press).

Given the tremendous economic disparities of different communities in Los Angeles, however, the patterns of settlement for these immigrants have not been even. Reflecting their lower socioeconomic backgrounds, most Latino immigrants settled in inner-city, low-income areas in search of relatively cheap housing. In Los Angeles, as in nearly all other U.S. cities, this meant moving into impoverished and underserved African American communities. There are now more Latinos than African Americans in historically African American communities such as South Central Los Angeles and the city of Compton (Adams & Williams, 1994; Chang, 1994, p. 103; Stewart, 1993). In addition, despite the attention given to the economic success of Asian Americans, the plurality of new Asian American immigrants settled in low-income, inner-city communities. In 1990, Koreatown and Filipinotown were homes to the largest concentration of Korean Americans and Filipino Americans in Los Angeles County (Ong & Azores, in press).

Although the scale of demographic changes in Los Angeles has been massive, political changes in the city have been just as profound. After a decade of protest in the 1960s, a clear national consensus emerged around racial politics. At the national level, under Democratic party leadership, the "Civil Rights Coalition" sought to increase the role of the government to boost the welfare state and to promote race-conscious policies designed to bring racial minorities into the mainstream. The Great Society programs of the Johnson administration raised and tightened the social safety net, whereas civil rights legislation and affirmative action policies attempted to redress past racial injustices. As Omi and Winant (1994, p. 106) have persuasively argued, the Civil Rights Coalition embodied an incrementalist approach to address racial injustice, *absorbing* those demands that could be accommodated within the existing political system and *insulating* out others that called for radical changes. Despite its incrementalist approach, the Civil Rights Coalition transformed the political scene in the United States, especially in those cities with large minority electorates. The victory of Tom Bradley over the recalcitrant Sam Yorty in 1973 marked a dramatic political transition in Los Angeles consonant with the nationwide political shift of the 1960s. As Sonenshein (1993, pp. 101-113) has pointed out, Bradley's victory represented a political triumph for a new coalition of African Americans and white liberals.

But although Bradley came into the mayor's office as a powerful symbol of racial equality and liberal reform, his legacy remains controversial at best, much like the broader, national legacy of the Civil Rights Coalition. Bradley received support from the moderates for his pro-growth, pro-business policies while appeasing white liberals and racial minorities by aggressively diversifying the racial and gender composition of the city government (Guerra, 1987; Sonenshein, 1993, pp. 151-155). At the same time, he received intense criticism from conservatives for his "tax-and-spend" social policies and "reverse racism" and from progressives for betraying the low-income, inner-city residents (Davis, 1990; Freer, 1994, pp. 185-186; Sonenshein, 1993). Yet overall, Bradley clearly reflected the broader political agenda of the Civil Rights Coalition. Although he brought in unprecedented numbers of racial minorities and women into the political system, he did so without initiating fundamental political or economic change.

Despite its modest and gradual approach, the Civil Rights Coalition came under attack, beginning in the mid-1970s. At the national level, "backlash politics" reached its height during the 1980s, with the landslide presidential victories of Ronald Reagan and George Bush. Although many factors contributed to this backlash, racial issues played a major role. This is especially true when one examines the crossover of "Reagan Democrats," the working-class whites who have traditionally been the mainstay of the Democratic electorate. During the 1970s, explicit racial issues such as busing, affirmative action, and minority set-aside programs and more implicit racial issues such as crime, social welfare, and immigration (both illegal and legal) led millions of white Democrats to become disaffected with the political agenda of the 1960s and pushed them to support conservative political leadership (Edsall & Edsall, 1991; Omi & Winant, 1994, chap. 7). In this way, racial politics served as the linchpin of the conservative political ascendancy, whose major domestic political policies were to slash the welfare state and to dismantle proactive racial policies and programs.

In California, the passage of Proposition 13 in 1978 and the victory of George Deukmejian over Tom Bradley in the gubernatorial race of 1982 clearly marked the shift in state politics to the right. In Los Angeles, the Civil Rights Coalition held off conservative challenges throughout the 1980s; given the tone of national and state politics, this was a tremendous political achievement by the Bradley coalition. The Civil Unrest of 1992, however, finally undermined Bradley's civil rights coalition. White liberals abandoned ship: Richard Riordan, a political novice, decisively defeated former city councilman Michael Woo, whose political ascendancy and agenda were closely linked with Bradley. As Sonenshein (1993) demonstrates, even in the 5th District—a predominantly white liberal district (West Los Angeles) that cast the decisive vote in Bradley's victory over Yorty in 1973—57% of the voters chose Riordan (pp. 109, 291).

With the apparent breakup of the Civil Rights Coalition in the aftermath of the Civil Unrest, racial politics in Los Angeles has now entered a state of enormous flux. In this time of uncertainty, two new coalitions have emerged, each defined largely by competing assessments over the failure of the Civil Rights Coalition and over the causes of the Civil Unrest. The first seeks to build a new multiracial coalition by espousing much of the conservative critique of the Civil Rights Coalition. The Establishment Coalition has articulated a political agenda based on reducing welfare spending, curbing crime, and aggressively fostering a probusiness environment. In contrast, the second also seeks to build a new multiracial coalition but on the progressive agenda of economic and racial justice. The Progressive Coalition seeks to bring together a broad range of constituents who have criticized the Bradley administration in the past for neglecting problems of economic inequality and racial tensions in the city (Regalado, 1994, pp. 216-220). Together, these two new coalitions have provided the fundamental division in Los Angeles politics since the Civil Unrest. Their differences have also defined political divisions within the various racial and ethnic communities in the city, and nowhere is that more evident than in the Korean American community.

The Civil Unrest has transformed the Korean American community. Politically, Korean Americans responded to the Civil Unrest with a clear conviction to participate more fully in the political process. Within weeks of the Civil Unrest, 30,000 Korean Americans marched through the streets of Koreatown to register their losses and collective grief (Park, 1994, p. 201). Korean Americans gave Civil Unrest their own name, *sa-i-ku,* literally 4-2-9, following the Korean tradition of naming key historical events after their dates (E. H. Kim, 1993, p. 216). The widespread use of the term reflects an effort to stress the significance of the Civil Unrest for the Korean American community: The term *sa-i-ku* seeks to give a historical weight and a social complexity to a key turning point in the Korean American experience, in a way that *riot* or *rebellion* cannot (Chang, 1994, p. 113). The widespread use of the term among Korean Americans also reflects areas of political agreement within the community regarding the Civil Unrest. Korean American political leaders unanimously felt that the Civil Unrest should serve as a political wake-up call for the community (Chang, 1994; Cho, 1993). They felt that the community's lack of political representation caused it to suffer the grossly disproportionate amount of economic damage. Korean American political leaders of all stripes have argued that mainstream political institutions, including the police, made conscious decisions to sacrifice the politically marginal and "disconnected" Koreatown as a buffer zone while scrambling to save more politically important and "connected" communities such as Hollywood, Westwood, and Little Tokyo (Chang, 1994, p. 113; Cho, 1993, pp. 201-202; Kang, 1993). In part, the widespread use of *sa-i-ku* among Korean Americans reflects an attempt to widen the meaning of the Civil Unrest, to include their own subjectivity, to claim a unique victimization, and to express a collective will to take action to prevent such victimization from happening again.

Similarly, the Civil Unrest has changed the basis of legitimacy for political leadership in the Korean American community in two important ways. First, before the Civil Unrest, many Korean American community leaders claimed and found their political base through their ties to "homeland" politics (Chang, 1988; I. Kim, 1981). Given the recent arrival of most Korean Americans, the real and symbolic ties to the Korean peninsula urged many Korean Americans to support various Korean political causes and to reproduce homeland politics within the Korean American community. Second, for those Korean American community leaders whose political base was rooted in U.S. and L.A. politics, political participation was limited to safeguarding rather narrow economic interests of Korean American small businesses (I. Kim, 1981; Min, 1995, Chap. 8). Groups such as the Korean American Grocers Association (KAGRO) and the Korean American Garment Contractors Association (KAGCA) used their considerable financial resources to make campaign contributions to gain access to incumbent politicians. But as business associations, they limited their political involvement to support probusiness policies. Because of their inability or unwillingness to articulate any specific political agenda aside from economic ones, many of the political gains were symbolic, such as winning the official designation of "Koreatown" from city hall. But despite this, Korean American business leaders had access to elected politicians and visibility in the community. Overall, the ethnic insularity of the Korean American community made both forms of leadership possible. Because these leaders readily recognized this, they had little interest in pushing the boundaries of political involvement beyond the narrowness of homeland politics or protecting economic interests.

Perhaps the most significant impact of the Civil Unrest on the Korean American community is the demise of this politics of ethnic insularity. As Korean Americans increasingly felt victimized by the political system in Los Angeles, the community began rallying around political figures who had ties beyond the confines of the ethnic community. Almost overnight, political leadership shifted. During the Civil Unrest, the nation *and* the Korean American community were introduced to a new group of community leaders: Angela Oh, Elaine Kim, Edward Chang, Bong Hwan Kim, and Jay Kim, all of whom could articulate the concerns of the Korean American community to a broader audience and thus reach political institutions outside the confines of the Korean American community (Chavez, 1994; Kang, 1993, p. B14; E. H. Kim, 1994; Park, 1994, pp. 199-201).

But although there was clear consensus in rejecting the politics of ethnic insularity, the community—like the rest of Los Angeles—has remained bitterly divided over how to participate most effectively in the politics of post-Civil Unrest Los Angeles. Although members implicated the politics of ethnic insularity to explain why the community suffered such an inordinate loss, Korean Americans have used their political efforts to support and provide leadership to *both* Establishment and Progressive Coalitions. In their participation, Korean Americans have pushed their interests into the city's political agenda and have found a new level of political visibility and involvement.

The New Multiracial Coalitions

THE ESTABLISHMENT COALITION:
KOREAN AMERICANS AND THE SHIFT TO THE RIGHT

During the primaries for the 1993 mayoral election, most of the leading candidates—Joel Wachs, Richard Katz, Linda Griego, J. Stanley Sanders, and Richard Riordan—ran on conservative platforms, whereas only two—Michael Woo and Nate Holden—demonstrated any sympathy for the liberal and progressive concerns (Sonenshein, 1993, pp. 286-287). After the field was narrowed down to two—Riordan and Woo—political alignments began to take shape that consolidated the Establishment Coalition. During the runoff election, Riordan, a venture capitalist who was well-known in L.A. political circles as a probusiness conservative, was able to win political support not only from the white moderates and the conservatives but also from key racial minorities. He was able to easily recruit conservative racial minority political leaders such as Sanders and Griego to the list of his supporters. To the surprise of some observers, however, Riordan also succeeded in recruiting more moderate and liberal minority politicians, including Yvonne Brathwaite Burke and Richard Alatorre, both Democrats and former members of the Bradley Coalition (Sonenshein, 1993, p. 290). In addition, Riordan turned to Jadine Nielsen—a moderate Chinese American Democrat from San Francisco—to manage his campaign and to aggressively seek support from the Asian American community. Riordan's message of probusiness fiscal conservatism and law and order appealed to many middle-class Korean Americans, including the large number of business associations that gave financial support to Riordan's campaign. In the end, 31% of Asian Americans cast their vote for Riordan (p. 293). By recruiting these highly visible and well-established racial minorities to his campaign, Riordan effectively answered one of the most critical questions of his leadership; a conservative, Republican Anglo can find political support from communities of color in Los Angeles.

The Establishment Coalition and its policies have come to dominate the agenda of the city since the Civil Unrest. This is especially true in the important struggle behind the politics of rebuilding. Rebuild Los Angeles (RLA) clearly embodies the approach of the Establishment Coalition. Originally established by Bradley, RLA remains the primary official response to the Civil Unrest of 1992. From its inception, its basic strategy for dealing with the profound economic and social problems called to attention by the Civil Unrest was to rally large private corporations to reinvest in inner-city communities and to create jobs. Reflecting this priority, RLA was initially dominated by corporate leadership

(Regalado, 1994, p. 207). But when it became clear that large-scale corporate investments were not forthcoming, RLA changed its focus to small-business development. To date, RLA has resisted funding programs that provide direct social services, insisting on the self-appointed and narrow mandate of providing political leadership and technical expertise to facilitate small-business development (Kang, Kim, Park, & Park, 1993; Regalado, 1994, pp. 206-207). In response to this position, critics have argued that RLA has spent enormous sums of money without accomplishing concrete and measurable outcomes (Dorono, 1993).

From the beginning, racial politics played a central role in shaping the leadership of RLA. Initially, Bradley appointed Peter V. Ueberroth as its head. Almost immediately, progressive racial minorities criticized the appointment of an Anglo who lives in Orange County to deal with the problems of the inner city of Los Angeles. To respond, Bradley and Ueberroth quickly diversified the leadership by creating a racially diverse executive committee and a board of directors. When Ueberroth left the chair, four cochairs took control: Bernard Kinsey, an Anglo; Linda Wong, a Chinese American; J. Stanley Sanders, an African American; and Tony Salazar, a Latino (Rainey & Feldman, 1993). In 1994, under pressure from Riordan, who grew impatient with charges of factionalism and unaccountability within RLA leadership, the board of directors appointed Griego to become the executive director (Regalado, 1994, p. 207).

Within the Korean American community, business interests and conservative Republicans joined to support the Establishment Coalition. As mentioned earlier, Korean American business interests attempted to stay away from partisan politics before the Civil Unrest, opting to stress political access. In part, this strategy made sense, given the generally conservative politics of the business interests and the dominance of liberal Democrats among the ranks of Los Angeles politicians. The Civil Unrest, however, undermined the faith that Korean American business interests had with these politicians. During the Civil Unrest, Korean American business interests felt neglected and abused, and in the rebuilding efforts, both parties have often found themselves on opposing sides on such contentious issues as the rebuilding of liquor stores in inner-city neighborhoods (Chang, 1994, pp. 112-114; Kang, 1994). As the Civil Rights Coalition lost its hold in city politics and a Republican mayor stepped into office, the Korean American business interests abandoned their nonpartisan politics and aggressively pursued their own political agenda (Park, 1994, pp. 209-210). At the same time that Korean American business leaders changed the strategy behind their political involvement, Korean American Republican activists stepped up their efforts to organize the community, launching the Korean American Republican Association (KARA), which has become one of the most visible political forces in the community (pp. 214-216). These Korean American Republicans argued that the Korean American community fell victim to incumbent politicians: Brian Choi, the general secretary of the KARA, states, "The riots changed my whole view of life. Even though Koreans have been giving money to political candidates, City Hall ignored us. It was a slap in the face" (Lope, Renwick, & Seo, 1993, p. B8).

Political campaigns further consolidated the alliances between Korean American Republicans and business interests. The successful campaign of Jay Kim to the House of

Representatives in 1992 represented the most significant victory for Korean Americans who support the Establishment Coalition. As a longtime activist in the ultraconservative faction of the Republican Party, Jay Kim was well versed in backlash racial politics and well-known in the Republican Party long before the Civil Unrest and the election of 1992. Indeed, his rise in Republican circles resulted in large part to his ultraconservative racial politics, including his vehement opposition to affirmative action and vigorous support for strict regulation of illegal immigration (Ward, 1992; Yi, 1993). As the only Asian American Republican in the U.S. Congress, Jay Kim has become a prominent figure in conservative politics, providing key leadership in both the campaign for Proposition 187 in California and the drafting of the Contract With America (Freedberg, 1994).

Within the Korean American community, Jay Kim has been celebrated across political lines precisely because his victory symbolizes the eclipse of ethnic insularity in Korean American politics. Ryan Song, the spokesperson for the Korean American Democratic Committee, argues that the election of Jay Kim "has tremendous symbolic value [because] he can play a major role in educating Korean-Americans about how the mainstream political process works and how we can be included" (Doherty, 1992, City Times sec., p. 11). Echoing a similar sentiment, Jerry Yu, the executive director of the nonpartisan Korean American Coalition, also congratulates Jay Kim for "show[ing] that Korean-Americans can play leadership roles not just in their own communities, but in the larger society too" (p. 11). Jay Kim's ultraconservative politics, however, have also invited harsh criticism from Korean American progressives who view his politics as a source of further interracial division and hostility. For instance, when Jay Kim openly expressed support for Proposition 187, Korean American progressives, including Angela Oh, Bong Hwan Kim, and Roy Hong, joined Latino leaders in condemning Jay Kim as "racist" and "anti-immigrant" (Sandalow, 1994, p. A3). Despite his critics, Jay Kim has built a powerful political base in the Korean American community.

Aside from electoral politics, substantive policy issues have also solidified Korean American support for the Establishment Coalition. The rebuilding of some 200 liquor stores destroyed during the Civil Unrest has been one such issue. At first, city hall saw an opportunity to respond to the decade-long criticism that too many liquor stores did business in inner-city communities (Chang, 1993; Freer, 1994). City hall imposed a conditional use variance process that would allow city hall, in consultation with local residents, to impose certain business practices such as restricting business hours and hiring security guards (Chavez, 1994). Because 175 of the 200 stores in question were owned by Korean Americans, the issue quickly became important for their community. The legal hurdles imposed by city hall effectively frustrated the rebuilding of liquor stores; 2 years after the Civil Unrest, only 10 of the 175 Korean American-owned liquor stores were rebuilt (Kang & Lacey, 1994). Throughout this process, Korean American liquor store owners have argued that they have been revictimized, this time, by liberal politicians. The Korean American Grocers Association—the organization representing the liquor store owners—found their greatest support from the Republicans who saw the liquor store issue as a key opportunity to win additional support among Korean Americans and to further alienate them from liberal Democrats.

In Sacramento, Paul Horcher, then Republican Assemblyman from East San Gabriel Valley, sponsored a bill (A.B. 1974) in 1993 that would remove the conditional use variance process in Los Angeles. Horcher phrased the issue as a case of liberal government intervening in the free market economy and a case of "African American racism" [sic] against Korean Americans (Kang, 1994, p. B3). After heated debate, the bill was defeated in the Local Government Committee of the Assembly. In a press conference that drew both the mainstream and Korean American media, Horcher, flanked by Korean Americans, blamed "African American politicians" for derailing the bill because "they want the Korean Americans out of their community" (Gladstone, 1994, p. B3). In an editorial published in the *Korea Times,* Shawn Steel and Michelle Park-Steel (1994) called on Korean Americans to "carefully assess who are their friends and who are their enemies" and to respond to the "legislative terror" unleashed on the community by the "Democrat[ic] left" (p. 3). Jerry Yu, the executive director of the nonpartisan Korean American Coalition, linked the issue more directly to the Bradley Coalition and the neglectful policies of the Civil Rights Coalition. He claimed that the "problems in South-Central were built up over the last 30 years" because of the indifference of "these local black politicians," who were now using the liquor store controversy to scapegoat Korean Americans for failed liberal policies in the inner city (Kang, 1994, p. B3). Although the defeat of A.B. 1974 clearly represented a loss for the Korean American Establishment Coalition, the effort won many Korean American converts, who now saw themselves as opposed to the politically powerful African American community and the "Democratic left."

THE PROGRESSIVE COALITION: BRIDGING RESISTANCE

In contrast to the Establishment Coalition, the Progressive Coalition sought to push interracial coalition building further to the left by espousing issues of economic and racial justice. Leadership for this effort came from community-based, grassroots organizations that have criticized Bradley in the past for placing too much emphasis on corporate interests at the expense of inner-city residents and for ignoring the concerns of working-class and poor minority communities. Members of the Progressive Coalition have also criticized the Bradley Coalition for failing to explicitly address the rising racial tensions between communities of color. For them, the Civil Unrest of 1992 was taken as the inevitable outcome of a long period of neglect of pressing problems of both economic inequality and racial tensions (Freer, 1994, p. 176).

The Progressive Coalition took shape on many fronts. First, existing ethnic-specific organizations either transformed themselves by undertaking projects that had multiracial relevance or diversified their staffs. For example, immediately after the Civil Unrest, the Coalition of Neighborhood Developers, a group of predominantly African American community development corporations, underscored the problem of poverty and the lack of social services in Koreatown. They hired a Korean American—Raphael Hong—to organize Korean American residents (Kang et al., 1993, p. 31). Similarly, Justice for

Janitors, a high-profile Latino labor union, hired Asian American organizers to organize their predominantly Latino rank and file (Roy Hong, interview, May 8, 1994).

Second, new multiethnic organizations have been founded to institutionalize the Progressive Coalition. The Latino Coalition for a New Los Angeles (the "Latino Coalition") and the Asian Pacific Americans for a New Los Angeles (APANLA) are groups that attempt to transcend specific ethnic boundaries to participate more effectively in the political process. The Latino Coalition was founded, in part, as a response to the backlash politics against the Latino community during and after the Civil Unrest: Federal agencies dispatched 1,000 Immigration and Naturalization Services and Border Patrol officers to Los Angeles and the surrounding cities, and they deported 700 undocumented residents. In response, a diverse group of organizations, including the Central American Refugee Center (CARACEN), the Mexican American Political Association (MAPA), and *La Unión de Comerciantes Latina y Afiladas,* formed the Latino Coalition in the summer of 1992 (Kang et al., 1993, p. 31).

In contrast, APANLA was initially founded in May 1992 to organize relief effort for Asian American communities. Many of its initial meetings focused on addressing the needs of the "victims' associations" that brought together mostly first-generation merchants and workers who had lost their businesses and jobs. Because most of the relief effort was organized along ethnic lines, however, with ethnic-specific institutions playing a major role, APANLA did not play a large part in providing relief. Instead, APANLA found its niche as the Asian American voice in the politics of rebuilding. Its membership includes diverse organizations ranging from those that are ethnic-specific to pan-Asian American, social service oriented to political advocacy oriented, and community based to professional. In a short time, APANLA has been effective; in Los Angeles, it has become the liaison between diverse Asian American communities and city hall. APANLA successfully campaigned for the appointments of Asian Americans to staff positions and to commissions. APANLA has also served as the link between the Asian American community and the federal government, and it has successfully leveraged important private institutions as well. APANLA pressured the *Los Angeles Times* to hire Asian American reporters, including K. Connie Kang, to cover the Korean American community (D. Ching, interview, May 5, 1994).

Finally, the Progressive Coalition has found new organizations explicitly aimed at making the Progressive Coalition *multiracial.* Perhaps the most important among them is the Multi-Cultural Collaborative (MCC). Founded in May 1993 and headed by a diverse group of cochairs—Cindy Choi, a Korean American; Ruben Lizardo, a Latino; and Gary Phillips, an African American—MCC espoused an explicitly progressive *political* agenda (C. Choi, interview, May 3, 1994; Regalado, 1994, pp. 226-227). The attempt to build a multiracial coalition on a shared political agenda marks a dramatic departure from previous efforts at multiracial coalition building because such coalitions in the past have usually avoided politics in favor of less contentious cultural and religious activities (Freer, 1994, p. 221). To date, however, MCC activities have largely centered on political activities and organizations that seek progressive solutions and a multiracial approach to community problems. The board of directors of MCC consists entirely of members of ethnic- or racial-specific organizations who not only have lent organizational and financial support to MCC but also have brought community-based political legitimacy to MCC's activities (C. Choi, interview, May 3, 1994).

Although political conservatives and business interests in the Korean American community have espoused the Establishment Coalition, progressive community activists and social service providers from that same community have provided support for the Progressive Coalition. In fact, in a short time, Korean American progressives have quickly coalesced into a significant political force in the city, largely because of their overlapping memberships in Korean American, Asian American, and multiracial organizations. Within the Korean American community, progressives have made their greatest impact by building their own faction within the Korean American Inter-Agency Council (KAIAC)—an umbrella organization of more than a dozen Korean American social service and political advocacy organizations (Kang et al., 1993, p. 30). Initially, KAIAC was formed after the Civil Unrest in an effort to limit the potential for organizational competition among Korean American social service agencies that could have jeopardized the rebuilding effort in the community. Since then, it has quickly become a major force in the community. To date, progressives within KAIAC have brought, for the first time, wide-ranging progressive issues to the community's political debate (H. S. Yang, interview, May 4, 1994). They have made issue of class equity in victims' assistance, the needs of immigrant workers, and the need for multiracial coalitions to address economic and racial justice. The progressives have won significant victories by winning the official endorsement of KAIAC and bringing broad political legitimacy behind these causes (Roy Hong, interview, May 8, 1994).

Outside their ethnic community, Korean American progressives have played a leading role in shaping key pan-Asian American and multiracial organizations. They helped establish both APANLA and MCC and, in so doing, clearly demonstrated their effectiveness in L.A. politics. In both instances, Angela Oh and Bong Hwan Kim provided key leadership (D. Ching, interview, May 5, 1994). Since then, other Korean American progressives, notably, Roy Hong (executive director of Korean Immigrant Workers Advocates) and Cindy Choi, have actively provided leadership and support to APANLA and MCC. Altogether, these Korean American progressives have ensured a central place for the Korean American community within the Progressive Coalition. This is important because in the past, left-leaning political movements in Los Angeles have been ambivalent or even hostile toward Korean Americans. Often, they saw no constituency in the Korean American community. Liberals and progressives readily assumed that most Korean Americans were either apolitical or conservative. As such, their political agenda did not address the concerns of the Korean American community; in turn, they failed to capture the community's political imagination and support. Without outreach, Korean American progressives were isolated and had little legitimacy to advocate progressive politics within the community. The ascendancy of Korean Americans, however, in addition to other Asian Americans, in the formation of a new Progressive Coalition holds the promise of changing both the political viability of the Progressive Coalition in Los Angeles and the politics of Asian American communities, including the Korean American community.

Conclusion: Lessons From Los Angeles

Many lessons can be drawn from the political developments in Los Angeles since the Civil Unrest of 1992. First, the experience in Los Angeles illustrates the continuing significance of race in U.S. politics. Both the Establishment and Progressive Coalitions have acknowledged the importance of racial inclusion and have explicitly sought to incorporate the different racial groups for building their political base and legitimacy. Although race is important, however, the profound demographic and social changes of the last two decades have made the politics of race much more complex. In Los Angeles, the biracial equation of the past has been eclipsed by the contemporary multiracial realities, whereby Asian Americans and Latinos are becoming increasingly important for determining political legitimacy and power. In the Establishment Coalition, the aggressive recruitment and the inclusion of racial minorities can be seen as the *defining difference* between the conservative coalitions forged by Riordan and the previous conservative coalitions in Los Angeles; this might be precisely why Riordan's coalition will prove to be more successful and durable than its predecessors. Clearly, Riordan has demonstrated that a conservative multiracial coalition can be forged in contemporary Los Angeles. The Establishment Coalition comes at a time when conservatives are once again rising to prominence in national and state politics, and it remains to be seen whether this multiracial lesson will be heeded by the conservatives this time around or whether they will continue their historical exclusion of racial minorities from their political vision and leadership (Omi & Winant, 1994, pp. 132-136).

For the Progressive Coalition, a multiracial coalition lies at the core of their new progressive politics. Although some have labeled this coalition the "Rainbow Coalition," drawing parallels to the coalitions built by Jesse Jackson during his two presidential campaigns, there are profound differences between the two (Sonenshein, 1993, p. 263). First, Latinos and Asian Americans were largely absent from positions of leadership in the Rainbow Coalition. Second, the Rainbow Coalition reflected an extension, and not a fundamental revision, of the Civil Rights Coalition; in its politics, the Rainbow Coalition stressed racial inclusion and economic growth but underplayed *interracial* inequality and conflict as well as *intraracial* class differences within racial minority groups. In this way, the Rainbow Coalition's main priority was to revitalize the Civil Rights Coalition by attempting to bring back white liberals and to inspire racial minorities back into the Democratic party (Omi & Winant, 1994, pp. 142-143). The Progressive Coalition in Los Angeles represents a sharp break, although as an opposition movement, its long-term viability has yet to be tested. But in its formation, the Progressive Coalition has created a new group of political leaders, institutions, and relationships that have changed the political landscape of Los Angeles.

Politics in the Korean American community clearly reflects the broad changes in Los Angeles ushered in by the Establishment and Progressive Coalitions. The politics of ethnic insularity has been replaced, and new leaders stress their ties to the competing coalitions. Republicans such as Jay Kim and Brian Choi, along with business interests such as KAGRO and KAGCA, urged the Korean American community to support the Establishment Coalition by stressing liberal indifference and hostility. Furthermore, they argued that the Korean American community—marked by economic success and social mobility—shares little with progressives who call for fundamental political changes and economic redistribution. Instead, Korean American members of the Establishment Coalition have appealed to the middle-class aspirations and anxieties in the Korean American community, and they have underscored their coalition's commitments to fostering probusiness environment, law and order, and "family values" (J. Kim, 1994; Steel & Park-Steel, 1994). In response, Korean American progressives claimed that the Korean American community has both a moral responsibility and the political interest to support the Progressive Coalition (Kwoh, Oh, & Kim, 1993; Oh, 1993). On an ideological front, they pointed to the historic link between progressive social movements—including the civil rights movement—and the removal of institutional barriers to economic mobility and political inclusion for racial minorities, including Korean Americans. In addition, progressives argued that the problems of racial conflict and inequality, all of which were made clear by the Civil Unrest, were the direct result of neglectful conservative policies under the Reagan and Bush administrations (Chang, 1994, pp. 101-102). Reflecting this assessment, Korean American progressives have charged that the Establishment Coalition under Riordan is misguided and divisive, and they have criticized other Korean Americans who have lent their support to the Establishment Coalition.

The current division in the Korean American community marks a turn in its politics. The political division reflects questions regarding *how,* and not *if,* Korean Americans should participate in the broader political process in Los Angeles. As the Establishment and the Progressive Coalitions compete for power in Los Angeles, divisions within the Korean American community have grown sharper. Clearly, it is too early to tell which of the coalitions will eventually establish dominance over the city and precisely how the Korean American community will contribute to that process. Remaining unequivocally true, however, is that neither Los Angeles nor the Korean American community will ever be the same.

References

Adams, E., & Williams, F. B. (1994, August 6). Protest over police beating continues. *Los Angeles Times,* p. B1.

Barringer, H. R., Gardner, R. W., & Levin, M. J. (1993). *Asian and Pacific Islanders in the United States.* New York: Russell Sage.

Chang, E. T. (1988). Korean community politics in Los Angeles: The impact of Kwangju Uprising. *Amerasia, 14,* 1, 51-67.

Chang, E. T. (1993). Jewish and Korean merchants in African American neighborhoods. *Amerasia, 19,* 2, 5-21.

Chang, E. T. (1994). America's first multiethnic "riots." In K. Aguilar-San Juan (Ed.), *The state of Asian America* (pp. 101-118). Boston: South End Press.

Chavez, L. (1994, August 28). Crossing the culture line. *Los Angeles Times Magazine,* 22.

Cho, S. K. (1993). Korean Americans vs. African Americans. In R. G. Williams (Ed.), *Reading Rodney King/Reading Urban Uprising* (pp. 196-211). New York: Routledge.

Davis, M. (1990). *City of quartz: Excavating the future of Los Angeles.* London: Verso.

Doherty, J. (1992, November 8). Korean Americans hail Kim's victory. *Los Angeles Times,* City Times sec., p. 11.

Dorono, A. (1993, August 29). United poor, middle classes can hold decision-makers accountable. *Los Angeles Times,* City Times sec., p. 23.

Edsall, T. B., & Edsall, M. D. (1991). *Chain reaction: The impact of race, rights, and taxes on American politics.* New York: Norton.

Freedberg, L. (1994, April 1). Divided we stand: The immigration backlash. *San Francisco Chronicle,* p. A1.

Freer, R. (1994). Black-Korean conflict. In M. Baldassare (Ed.), *The Los Angeles riots* (pp. 175-204). Boulder, CO: Westview.

Gladstone, M. (1994, August 30). Bill to ease rebuilding of Korean American stores fails. *Los Angeles Times,* p. B3.

Guerra, F. J. (1987). Ethnic office holders in Los Angeles County. *Sociology and Social Research, 71,* 89-94.

Hacker, A. (1992). *Two nations: Black and white, separate, hostile, unequal.* New York: Scribner.

Hing, B. O. (1993). *Making and remaking Asian America through immigration policy, 1850-1990.* Stanford, CA: Stanford University Press.

Kang, K. C. (1993, May 29). Asian-Americans seek role in L.A. renewal. *Los Angeles Times,* pp. B3, B14.

Kang, K. C. (1994, July 21). Store owners fight restrictions on reopening. *Los Angeles Times,* p. B3.

Kang, K. C., & Lacey, M. (1994, July 15). Court rejects appeal of rules for liquor stores. *Los Angeles Times,* p. B1.

Kang, M., Kim, J. J., Park, E. J. W., & Park, H. W. (1993). *Bridge toward unity.* Los Angeles: Korean Immigrant Workers Advocates.

Kim, E. H. (1993). Home is where the han is. In R. G. Williams (Ed.), *Reading Rodney King/Reading urban uprising* (pp. 215-335). New York: Routledge.

Kim, E. H. (1994). Between black and white. In K. Aguilar-San Juan (Ed.), *The state of Asian America* (pp. 71-100). Boston: South End Press.

Kim, I. (1981). *New urban immigrants: The Korean community in New York.* Princeton, NJ: Princeton University Press.

Kim, J. (1994, December 28). Debate on "Contract With America" [Radio program]. Los Angeles: Radio Korea.

Kwoh, S., Oh, A. E., & Kim, B. H. (1993, May 3). Don't let up now that verdicts are in. *Los Angeles Times,* p. B7.

Lope, R., Renwick, L., & Seo, D. (1993, May 28). Charting a new course. *Los Angeles Times,* p. B8.

Lyman, S. (1974). *Chinese Americans.* New York: Random House.

Marable, M. (1991). *Race, reform, and rebellion.* Jackson: University Press of Mississippi.

Min, P. G. (Ed.). (1995). *Asian Americans.* Thousand Oaks, CA: Sage.

Munoz, C. (1989). *Youth, identity, power.* London: Verso.

Oh, A. E. (1993). Rebuilding Los Angeles: Why I did not join RLA. *Amerasia, 19,* 157-160.

Oliver, M. L., Johnson, J. H., & Farrell, W. C. (1993). Anatomy of a rebellion. In R. Williams (Ed.), *Reading Rodney King/Reading urban uprising* (pp. 117-141). New York: Routledge.

Omi, M., & Winant, H. (1994). *Racial formations in the United States* (2nd ed.). New York: Routledge.

Ong, P., & Azores, T. (forthcoming). Asian immigrants in Los Angeles: Diversity and divisions. In *Struggles for a place.*

Park, W. (1994). Political mobilization of the Korean American community. In G. O. Totten & H. E. Schockman (Eds.), *Community in crisis* (pp. 199-220). Los Angeles: University of Southern California, Center for Multiethnic and Transnational Studies.

Rainey, J., & Feldman, P. (1993, November 12). New RLA chief likely to be corporate leader. *Los Angeles Times,* p. B1.

Regalado, J. A. (1994). Community coalition-building. In M. Baldassare (Ed.), *The Los Angeles riots* (pp. 205-236). Boulder, CO: Westview.

Reimers, D. M. (1985). *Still the golden door.* New York: Columbia University Press.

Sandalow, M. (1994, April 14). GOP defends immigration plan. *San Francisco Chronicle,* p. A3.

Sonenshein, R. J. (1993). *Politics in black and white: Race and power in Los Angeles.* Princeton, NJ: Princeton University Press.

Steel, S., & Park-Steel, M. E. J. (1994, September 7). Outcome of AB 1974: Korean-Americans strangled again. *Korea Times: English Edition,* p. 3.

Stewart, E. (1993). Communication between African Americans and Korean Americans. *Amerasia, 19,* 23-54.

Ward, M. (1992, June 21). Local elections/41st Congressional District. *Los Angeles Times,* p. B3.

Yi, D. (1993, October 6). From NAFTA to immigration: Rep. Kim speaks out before KA Republicans. *Korea Times: English Edition,* p. 1.

Payment

ELOISE KLEIN HEALY
Poet

French poodles are missing.
Garbage cans get spilled
where condominium tracts
pave over animal trails.
They're setting traps in the foothills
where a coyote carried off a child.
Wilderness not won over creeps back in meaner.

I call it compensation for natural disaster
when flames rise out of Fairfax Avenue
where the city sits on pockets of tar,
when spindly canyon houses and their stilts
wash down after the rains
and mansions explode in the surf at Malibu.

The newly mapped slip-fault
undergirding the high-rise Civic Center testifies
this power will return,
repossessing and collecting its dues.

SOURCE: "Payment," by Eloise Klein Healy, from *Artemis in Echo Park,* by Eloise Klein Healy. Ithaca, NY: Firebrand Books. Copyright © 1991 by Eloise Klein Healy.

Multiracial Organizing Among Environmental Justice Activists in Los Angeles

10

LAURA PULIDO

Introduction

The physical geography of Southern California and its manufacturing industries generate an array of environmental and natural resource problems, including the most severe air pollution in the nation. Specifically, the results of automobile dependency and intense industrialization are exacerbated by the surrounding mountains and a thermal inversion layer, which produce both smog and air toxins.[1] Despite a number of attempts on the part of various municipalities to clean up the air (Lents & Kelly, 1993), real progress was not possible until the establishment of a regional entity (FitzSimmons & Gottlieb, 1993), the South Coast Air Quality Management District (SCAQMD). The agency was charged with attaining national ambient air quality standards for criteria pollutants by the year 2010 through a comprehensive plan (SCAQMD, 1989; see also Bloch & Keil, 1991).[2] The plan has since been discredited by diverse constituencies; for example, environmentalists successfully sued the agency, arguing that the plan was not sufficiently strong to bring

AUTHOR'S NOTE: Many thanks to Michael Dear for his useful comments. I am responsible for all interpretations and shortcomings. The data for this chapter are based on numerous years of participant/observation in the local environmental justice moment and a set of interviews conducted in 1993. Given the lag time between the reported political work and the publication date, this analysis should be seen as a "snapshot" of the early 1990s.

the region into compliance, and business interests have waged an intense battle to erode the agency's powers (SCAQMD, 1992). It has nonetheless succeeded in raising the general level of consciousness and concern about air pollution.

One set of communities that has become involved at an unprecedented level in air pollution and other environmental matters is low-income and minority groups,[3] which previously had little formal engagement in environmentalism.[4] This rise in concern and action was partly because air pollution became a regular component of the public consciousness and discourse but also because of the growing awareness of environmental racism and the environmental justice movement. *Environmental racism* denotes non-whites' disproportionate exposure to various forms of pollution and environmental hazards (Bullard, 1990; Citizens for a Better Environment, 1989; United Church of Christ, 1987; U.S. General Accounting Office, 1983); *environmental justice* is the name of the movement that has arisen to counter these inequities. Given L.A.'s industrial base, its large nonwhite population,[5] and its entrenched patterns of residential segregation, it is not surprising that minorities in Los Angeles are disproportionately exposed to uncontrolled hazardous waste sites (United Church of Christ, 1987), facilities emitting air toxins (Burke, 1993), and attempts to locate incinerators (Russell, 1989). Accordingly, Los Angeles has been the site of significant environmental justice activism. Three organizations in particular, all of whom have organized around various forms of air pollution, have become local and national icons of the environmental justice movement. They include the largely African American Concerned Citizens of South Central Los Angeles (CCOSCLA), the *Mexicana* Mothers of East Los Angeles (*las Madres;* MELA), and the multiracial Labor/Community Strategy Center (LCSC). Importantly, each organization has mobilized in a community burdened with high levels of air pollution, South Central, East L.A., and Wilmington, respectively. Although each group has its own issues, constituencies, and problems, they often work in coalition and are considered exemplars of how environmental issues, in particular, environmental racism, have the power to bring diverse groups together. "The movement for environmental justice . . . mobilize[s] community-wide coalitions built across race, ethnic, and class lines" (Taylor, 1992, p. 44).

Indeed, the ability to transcend lines of difference has been a hallmark of the movement. Although the environmental justice movement includes many white communities across the nation (Szasz, 1994; see also Hofrichter, 1993), nonwhites have organized around the frame of environmental racism (see Capek, 1993) and have built on the idea of race in the formation of a movement and related identities. The environmental racism frame has served to bring thousands of nonwhites into the environmental fold by employing the umbrella identity, people of color, which in turn allows the possibility of multiethnic activism. This is important because diverse groups cannot engage in collective action unless they recognize some commonalities and develop at least a partial collective identity. By emphasizing that people of color are disproportionately affected by pollution, racial inequality (racism) has become the base of a new, proactive identity, one that is a unified and hegemonic racial representation. Organizing within the framework of antiracism should be seen as a strategic move by marginalized agents because it enables disparate groups with limited power to consolidate their efforts and to become more powerful. The possibility of a unitary identity occurs not only because African Americans and Latinos, for example, confront similar environmental conditions and experience racism and economic subordination but also because these commonalities are clearly articulated by movement leaders. Nevertheless, it is still unclear how racial identities are being constructed and/or destabilized by rank-and-file environmental justice activists themselves. Because there appears to be a mismatch between the discourses of environmental justice leaders and the larger memberships, the political consequences of this situation for both minority communities and the environment need to be considered.

I argue that the multiracial rhetoric and discourse of environmental justice activists in Los Angeles are much stronger than the reality because activists' "old" racial identities are highly resistant to change. Despite the portrayal of environmental justice groups in Los Angeles as on the forefront of multiracial activism, the truth is that a panracial identity is quite limited, and when it does occur, it is highly contained. One reason that a collective minority identity and activism have not trickled down to the larger membership is that specific ethnic groups experience racism differently within Los Angeles, causing them to cling to more unified and less ambiguous identities that they feel will better meet their political objectives. In contrast, the adoption of a new panracial identity appears to be much stronger among middle-class and professional nonwhites, who are better able to collectively posit their needs and those of their sponsors, while serving as a voice for all nonwhites on environmental matters. This is an important example of how even within the same racial/ethnic group, class may serve to create different forms of racialization and racism.

Questions of identity are crucial to understanding the efficacy of a social movement. Recent scholarship not only has focused on the contingent nature of identity but, more specifically, has sought to reveal the fragmentation and multiplicity of identities. In contrast, less attention has been directed to those instances when individuals and groups cleave to more fixed identities. This chapter attempts to illuminate some of the complexities surrounding collective action among historically excluded groups; specifically, it asks, to what extent have environmental justice activists forged a multiethnic movement? Why are some nonwhites hesitant to embrace a broader identity, one that will, perhaps, assist them in their quest for environmental justice? Why is this not the case for more privileged nonwhite activists?

To address these questions, I will first provide an overview of environmental justice activism in Los Angeles. Second, I will examine the idea and reality of multiracial organizing among environmental justice activists, paying special attention to the obstacles that exist to building a multiracial movement among working-class African Americans and Latinos.[6] Third, I will consider why multiracial organizing has proved to be such a challenge among working-class nonwhites but less so among privileged minorities. I will conclude with some of the broader implications of these findings.

Environmental Justice Activism in Los Angeles

Several writers have traced the awareness of environmental racism to an effort in 1982 to dump polychlorinated biphenyl (PCB)-contaminated soil in a largely African American community in North Carolina (Lee, 1992). About this time, a growing number of communities across the nation were finding themselves confronting their own local versions of Love Canal, the upstate New York neighborhood built over an abandoned hazardous waste site. As these communities realized the widespread nature of the problem, they began to coalesce through informal networks and coalitions to assist each other. I emphasize again that affected were not only minority communities but rather communities of all races, ethnicities, and incomes (Brown & Masterson-Allen, 1994; Szasz, 1994). Nevertheless, because of the high frequency with which low-income and minority communities were affected and their unprecedented activism, a specific frame of environmental *racism* was developed (McGurty, 1995). As nonwhite communities found themselves on the receiving end of an undesirable project or suffering from a particular environmental hazard, the situation was interpreted as one of environmental racism.

Thus, when a waste-to-energy incinerator was proposed for the largely African American community of South Central Los Angeles in 1983, it was seen as a clear case of environmental racism. The incinerator, known as the LANCER project,[7] was the city's response to L.A.'s growing solid waste problem in the face of opposition to landfill expansion. Only by accident and late in the game did local residents discover what was planned for them. As word spread, neighbors organized themselves to form Concerned Citizens of South Central Los Angeles (CCOSCLA). On learning what the project would entail, residents were both shocked and outraged, not only because the incinerator would be in close proximity to schools and homes but also because the high levels of dioxin, heavy metals, and fly ash would compound an already serious air pollution problem in South Central. In the words of Shiela Cannon, an outspoken foe of the project, "They are trying to kill us" (quoted in Russell, 1989, p. 26). The activists, primarily working-class African American women, soon became experts on most facets of incineration, from the permit process to end-state emissions. Armed with this information and sheer determination, they launched a multiyear battle against the city of Los Angeles. Two of their most successful weapons were in soliciting support from Anglo Westside environmental groups, such as Not Yet New York and Greenpeace, and in having a new environmental impact report (EIR) drawn up by UCLA faculty and students that effectively challenged the legitimacy of the previous document. Ultimately, CCOSCLA prevailed, and the incinerator was defeated.

During and subsequent to the battle, CCOSCLA attracted a great deal of media attention both locally and nationally. Not only did they successfully resist the incinerator, but also they became the impetus behind the city's recycling program. Their commitment to community well-being and proactive environmental agenda resulted in a great deal of moral authority. Consequently, CCOSCLA became one of the symbols of the environmental justice movement. After the conclusion of the incinerator battle, CCOSCLA maintained its organizational structure and went on to address a variety of other issues, including lead poisoning, housing, and economic development. To date, the organization has initiated eight affordable housing projects and two economic development ventures while continuing to have a voice in environmental justice matters.

Soon after LANCER was proposed, a similar project was announced in 1985 for the city of Vernon. Instead of burning solid waste, however, the Vernon incinerator was intended to burn hazardous waste. Vernon is a small city adjacent to East Los Angeles and is purely industrial, giving it the ranking of the dirtiest zip code in California (Kay, 1994, p. 156). Aside from the large concentration of polluting industry clustered along the Santa Ana freeway and the Southern Pacific railroad, the wind patterns are such that heavily Latino East Los Angeles has some of the highest concentrations of air toxins and smog in the South Coast Air Basin. MELA, which was formed previous to the incinerator announcement (Pardo, 1990), opposed the project with a vengeance. As in South Central, there was great concern over the proximity of the incinerator to schools, housing, and food-processing plants. But most incredible of all, the California Department of Health Services did not require that an EIR be conducted. In response, MELA mobilized large numbers of people to engage in weekly protests but also pursued a litigation strategy in conjunction with the Natural Resources Defense Council and the city of Los Angeles. Eventually, as community opposition mounted, the demand for an EIR grew, and, as a history of pollution violations was revealed, California Thermal Treatment Service withdrew its bid to build an incinerator.

MELA attracted attention from as far away as England, India, and Malaysia and quickly rose to prominence in the environmental justice movement. MELA has continued to participate in environmental issues, such as its water conservation program (which provides a funding base for other projects), but has also broadened its scope to include other initiatives such as scholarship, immunization, and antigraffiti programs.

Something that caught the attention of many people both in Los Angeles and nationally was the way CCOSCLA actively assisted MELA, sharing what information and skills they had learned along the way.

> In perhaps the most notable success story, such a coalition, forged by the Concerned Citizens of South-Central Los Angeles, succeeded in scrapping plans to build a huge garbage incinerator in a predominantly African American, inner-city neighborhood. Today African Americans from CCOSCLA are joining forces with the Hispanic members of MELA, who in turn are forming a coalition with Chinese residents of nearby Lincoln Heights. (Russell, 1989, pp. 23-24)

This was the beginning of the multiracial interaction.

The Labor/Community Strategy Center (LCSC) differs from both CCOSCLA and MELA in that it was not organically formed in response to a place-specific threat; rather, it is a response to ambient air pollution, which is far more difficult to mobilize around.

The Center began with labor activists who felt the environment was an important vehicle for organizing and saw the need for an environmental vision to counter more mainstream and elitist approaches (Labor/Community Strategy Center, 1989; Mann, 1991). The Center is seeking to build a new left coalition in Los Angeles, and its politics are explicitly multiracial and anticorporate. Focusing on the pollution generated by corporations and their inordinate influence over local air quality policy, the Center is trying to expose the various lines of power and difference woven into the production, exposure, and regulation of air pollution. Emerging in the late 1980s, the Center, and specifically its environmental project known as the Watchdog, formed partnerships with key figures of MELA and CCOSCLA and began making its presence known at SCAQMD meetings. Its objective was to shift the discourse of the SCAQMD from one of catering to business to one that put public health first. In addition, the Watchdog began canvassing in low-income minority neighborhoods to generate interest in both air pollution and left politics—hoping people would see the connection. This strategy was not always successful. Nevertheless, during its early years, the Watchdog was able to influence policy, albeit to a limited degree. For example, it fought for stricter limits on toxic air emissions, strove to ensure that policy did not disproportionately burden low-income and minority communities, insisted on full translation in all SCAQMD proceedings, and tried to force industry to assume its share of responsibility. The Watchdog's trajectory took a sharp turn in the 1990s, however, as Los Angeles was mired in the worst recession since the Great Depression. Industry took this opportunity to lobby hard against the SCAQMD, which it blamed for Southern California's economic crisis.

But winning on SCAQMD turf had never been the principal goal of the LCSC. Instead, its ultimate objective was to build a movement to wage a campaign against a major polluter (as part of building a left consciousness and politics). Only by targeting a large emitter, the Watchdog argued, would people appreciate the immense power of corporations; equally important, it would provide a clear organizing focus. Because of the heavy concentration of the petrochemical industry in the South Bay, the area is one of the most significant clusters of air toxins in L.A. County (U.S. Environmental Protection Agency, 1994). For this reason, the Watchdog chose to focus on the hardscrabble, heavily Latino (but ethnically mixed) community of Wilmington, which sits in the shadow of several refineries. For several years, the Watchdog organized in the Wilmington area, trying to build a core of local leadership, while also working to build a countywide movement to boycott Texaco. The Watchdog believed that only by building a regional movement would local communities have sufficient strength and resources to challenge polluters. Texaco was the chosen target not only because it is a major polluter[8] but also because it has a poor record of regulatory compliance and operates a large number of local stations (thus facilitating a boycott). Texaco's role as a target was solidified in October 1992 when there was a major explosion at the refinery, resulting in widespread illness and concern. The results of the boycott have thus far been mixed, as the Watchdog has found it difficult to inspire low-income people to become involved in air pollution issues, despite the fact that they breathe and smell noxious fumes. It has had no problem, however, attracting many Anglos and middle-class individuals who are drawn by its politics. The hard work, political vision, and multiracial nature of the center have caused it to become a key fixture of the environmental justice movement.

Although there are often coalitions that include other organizations, such as Greenpeace and Citizens for a Better Environment,[9] CCOSCLA, MELA, and LCSC pose the strongest examples of multiracial environmental justice work. Not only are they composed largely of people of color, but they are known in activist circles throughout the country. Regionally, they participate in the Southwest Network for Economic and Environmental Justice (SNEEJ), and locally they are considered key players in environmental justice matters, particularly when a minority or environmental justice viewpoint is desired. For instance, when the SCAQMD established an ethnic advisory council, each group was invited to participate.

MULTIRACIAL ENVIRONMENTAL JUSTICE ACTIVISM IN LOS ANGELES

Groups and associations of black and brown people are waging grassroots environmental campaigns all over the country. . . . The activity is indicative of a minority grassroots movement that occupies a distinctive position relative to both the mainstream movement and the white grassroots environmental movement. The minority movement is anti-bourgeois and anti-racist. *It capitalizes on the social and cultural differences of people of color* as it cautiously builds alliances with whites and persons of the middle class. (Austin & Schill, 1991, p. 71; emphasis added)

Environmental racism has served as a catalyst in creating cooperative interaction among nonwhites. Ben Chavis, a national leader of the environmental justice movement, has voiced an explicitly multiracial agenda: "It is our intention to build an effective multiracial, inclusive environmental movement with the capacity to transform the political landscape of this nation" (quoted in Kerr & Lee, 1993, p. 10). Multiracial coalitions are nonetheless largely the actions of organizations' leaders, who often are more politically developed and morally committed to fostering multiracial interaction and see the benefit of presenting a united front. But this is not necessarily the case among the larger membership of the movement.

One of the lead organizers of the LCSC, an African American, summed up his commitment to a multiracial movement but acknowledged its nascent nature:

Building a multiracial movement is critical because the toxic problem is so pervasive. They're everywhere and people are endangered. . . . The reason we have to be together is because the enemy . . . concentrated corporate wealth, is so formidable [that] . . . no ethnic group, Blacks, Latinos, Asians are large enough or powerful enough by itself to make a dent in it. . . . [But] we are not a movement yet. There is movement towards a movement. We have the beginnings of a multiracial organization in this group. It's the only game in town. And increasingly, I'm told by people, the only one in the country. The best multiracial group in the country! That to me is frightening. We have a lot of work to do and I don't know if I'll see it in my lifetime.

Although the Watchdog is explicitly a multiracial organization trying to build a multira-cial movement, it has run into difficulties, particularly in recruiting African Americans. Despite the center's board, staff, and organizers "looking like the civil service" according to one member—meaning it is characterized by great racial and ethnic diversity—and the close working relations of the center with CCOSCLA and MELA, the work of building a movement has been far more challenging. Trying to mount a countywide initiative, the Watchdog initially went door-to-door in largely African American sections of South Central Los Angeles, but eventually this work ceased because of a lack of interest as well as a shift in attention to Wilmington. In many ways, Wilmington was an ideal community in which to organize because although it was largely Latino, it had a significant African American population and is adjacent to Carson, an amalgam of middle-class African Americans, Asian Americans, and Anglos. This portion of Los Angeles is heavily affected by an array of air pollution sources, leading the Watchdog to believe that a real potential for building a multiracial movement existed. For several years, organizers went door-to-door, spoke at churches, and worked with youth at the local high school but were not able to create a multiracial organization. Instead, it wound up being overwhelmingly composed of Latino immigrants with a sprinkling of Filipinos, largely from Banning High.

One reason for this is language. The Spanish language plays a highly contradictory role in the creation of a diverse movement and offers important insights into how boundaries are created and into the meanings embedded in them. Although environmental justice supporters are quick to applaud the use of "local forms of culture," language can actually be a dual-edged sword in organizing. "For poor minority folks, social and cultural differences like language are not handicaps, but the communal resources that facilitate mobilization around issues like toxic poisoning" (Austin & Schill, 1991, p. 74). This is only partially true. The Center has a large number of bilingual staff, including one African American organizer fluent in Spanish. Because Los Angeles County is 40% Latino, any effective mobilization will necessarily have large Latino participation and use bilingual materials and strategies. In the course of organizing in Wilmington, the Watchdog adopted the practice of conducting meetings in the majority language. Thus, if monolingual Spanish speakers are predominant, the meetings are held in Spanish with translation services provided to those who are monolingual English and vice versa. Although such a strategy is cumbersome and time-consuming, it is necessary to allow Latino immigrants to participate and is evidence of the Center's commitment to building a diverse constitu-ency. But at the same time, language has been an obstacle to the participation of African Americans, who in the words of one organizer, "perceive the Watchdog as a Latino group because we offer translation." She noted that when African Americans did attend, they did not appear to be comfortable with multiple languages, and some voiced the opinion that the meetings should be held in English. In this case, a bilingual organizational structure that allows outreach to one oppressed group has not enhanced the participation of another.

The Spanish language is also noteworthy because of what it signifies. In contemporary Los Angeles, language ability has become a partial signifier of one's politics and position. For example, among bilingual and English-speaking activists, there appears to be a clear commitment to building a multiracial movement. This can be seen in the growing number of Anglos who have learned Spanish—or at least try (far more Anglos speak Spanish than do African Americans or Asians). At a more fundamental level, the ability to speak English, either as a native or adopted tongue, demonstrates a familiarity with the dominant Anglo society and its ideals, particularly the concepts of people of color, minorities, and multiracial activism. Such constructs simply do not exist in Latin America. The notion of multiracialism is quite foreign to many immigrants, and some are downright hostile toward African Americans. Some Latino immigrants, for example, explained that Mexican television (much of it from the United States) consistently portrays African Americans as dangerous and without morals; for this reason, a number of Latino immigrants avoid them. In more than one instance, Latino immigrants would not attend a meeting held in what they felt was a largely African American neighborhood "because you never know what those people will do."

The situation of single-ethnicity organizations offers its own set of problems and insights. For CCOSCLA, language once again loomed as a major issue. CCOSCLA is known as an African American organization, and South Central is often depicted as the heart of black Los Angeles. The truth is, however, that South Central has undergone a dramatic demographic shift, and parts of South Central are now more than 50% Latino. Nevertheless, CCOSCLA has an all-African American board (17 members), one Latina staffperson, and an African American youth organizer who speaks Spanish. The single Latina felt she was responsible for cultural as well as language translation but did not feel all segments of the organization were committed to including Latinos.

> What I wanted to accomplish when I got into the job was to diversify the organization and to help the crossover to a more multicultural environment, which I have not been successful in doing, which *they* have not been successful in doing. . . . [But] there's that certain acknowledgment that there's a need to draw Latinos into the organization, because that community is now becoming 60% Latino.

In recent years, CCOSCLA has turned its attention to childhood lead poisoning—a neglected issue of grave concern to both African American and Latino urban communities. Its strategy, however, based on organizing through local "block clubs," has not attracted strong Latino participation. Some interpret this as a lack of interest and passivity by Latinos; others have pointed out that Latinos prefer to organize through the church or the workplace. Some have charged that the organization does not truly welcome Latinos. Whatever the case, Latino participation has been slight, although some recent improvements have been attributed to Latinos "getting used to Blacks." One African American board member of CCOSCLA observed that translation can serve as an effective and inclusive organizing tool:

> We get some to come to the meeting and get them involved. "Can you translate?" Now they feel that they got a part. So next meeting they be bringing their families so they can say "Watch me! See what I can do!"

Nevertheless, numerous members of CCOSCLA acknowledged there were difficulties with Latino outreach and African American/Latino prejudices, but they were doing their best to resolve them.

> One of the barriers is that people have misconceptions about each other, all of these negative stereotypes we are fed by the media. So one of the first things we do at a lot of Block Clubs is to get those and lay them straight on the table. . . . One of the first things that Latino people say is that they thought we did most of the crime. . . . So when they get here they find that they are exploited by mostly other than us. So then we say, "Had this all really been true, you would not have been able to move into our communities because we would have made it difficult."

The situation is further complicated by the larger politics of language engulfing South Central. There is a general tension surrounding monolingual Latino immigrants and African Americans regarding education and language. Because of the large influx of Latino immigrants, many South Central schools must now offer bilingual instruction. From the perspective of African Americans, resources that once were available for music, art, and recreation programs have now been shifted to bilingual programs. Moreover, some African Americans complained that although Latino children were being taught two languages (Spanish and English), African American children were receiving instruction in only one. Some parents wanted their children to learn Spanish; others were disturbed that a large percentage of instruction was conducted in a language their child did not understand; still others simply opposed the diversion of resources.

For all these reasons, language has become one of the most meaningful forms of difference in multiethnic Los Angeles. The Spanish language has become a primary but complex boundary, not just between African Americans and Latino immigrants but also between immigrants and native Latinos and between Anglos and immigrants. Language is at once a point of difference and a site of contestation, particularly because Spanish has become racialized. For Latinos, language has became one of the most important forms of social difference, one that is subject to both institutional and informal forms of racism. For this reason, a major battle between the SCAQMD and environmental justice activists erupted over how translation was to be conducted. Although activists felt simultaneous translation (headphones) would be most inclusive, the SCAQMD and business interests advocated isolating all Spanish speakers in a separate room, a move that activists interpreted as highly racist. Clearly, any attempt to forge socially just and inclusive environmental policy must inevitably rest on a wide acceptance of bilingualism/translation and on making available the necessary resources to teach English to large numbers of people.

Language played a lesser role in accounting for the limited interactions between CCOSCLA and MELA. The Spanish language barrier is more permeable between the communities of South Central and East Los Angeles, largely because the latter community has a greater percentage of Mexican Americans who have deeper roots in Los Angeles and are therefore more familiar with the English language and dominant American norms. Moreover, there is a history of interaction between the two groups as they have marched in solidarity with each and readily shared strategy and tactics. One leader of CCOSCLA explained their initial contact:

> I had seen Juana and Ricardo, but I didn't know them. Then all of sudden we find out they were fighting an incinerator, and we were fighting an incinerator. So we found a lot of commonalities. . . . We sat down to talk and developed some real relationships, talked about how we can work together.

But subsequent to these mobilizations, contact largely ceased except among the leadership.

Part of the problem is the racialization of space. Because the geography of Los Angeles is produced by social relations, racism plays an inherent part in such processes. South Central is consistently portrayed to the world as a dangerous, dark place. It is inscribed with blackness. By the same token, East Los Angeles has traditionally been home to the region's large Mexican-origin population; it too is considered crime-ridden, poor, and undesirable. And because they have traditionally been racially segregated, it is quite possible for Eastside Mexicanos to go almost their entire lives without really knowing an African American and vice versa (although such is no longer the case in South Central). There can be no denying that numerous Eastside residents harbor varying degrees of prejudice against African Americans. One member of MELA offered some insight into how she interpreted what she considered to be African Americans' hostility: "I can understand why they're bitter, but I didn't cause them harm!" Nor did she perceive African Americans to be warm, open, or cooperative people. This is an example of how the process of politicization in one arena (the environment) has not led to a broader political consciousness despite the efforts of leaders to articulate such linkages. For both Eastside Latinos and South Central African Americans, it is a far trip geographically and psychologically to each other's spaces and lives. The social and spatial boundaries are simply too difficult to cross after the urgency of a campaign has subsided. Although both communities share the same air basin and confront similar economic, environmental, and social problems, a common identity is lacking. Only at specific moments can they unite as people of color, despite the hard work and political vision of their leaders.

Emergent/Resistant Identities in Los Angeles

Why have grassroots activists been so hesitant to forge a multiracial movement? What does it teach about identity? How has it influenced environmental politics? These questions can best be answered by looking to recent theoretical developments in identity politics.

We all have multiple, ever changing identities. The destabilization of a fixed identity is arguably one of the greatest contributions of postmodern thought to the fields of identity politics and social movements. That social scientists have for so long focused on a single identity and deemed only certain types of social movements as legitimate is clearly an artifact of modernity, and, as a consequence, such analyses may never have properly captured the reality of people the world over, especially of marginalized persons (Nash, 1992).

These insights have been particularly useful in leading to a reassessment of race and ethnicity (Brah, 1992; Gilroy, 1993; Gregory, 1993; Jackson, 1987). Within this literature, one of the most important developments is the move away from unified hegemonic racial identities. Scholars and artists have drawn attention to the problematic nature of our refusal to acknowledge diversity *within* a particular racial group. Writing on the situation of blacks in England, Gilroy (1993) notes the tension between those who subscribe to a unitary racial subject and those who do not.

> Its [traditional anti-racist activists] absolutist conception of ethnic cultures can be identified by the way in which it registers uncomprehending disappointment with the actual cultural choices and patterns of the mass of black people in this country.... This perspective currently confronts a pluralistic position which affirms blackness as an open signifier and seeks to celebrate complex representations of a black particularity that is *internally* divided: by class, sexuality, gender, age and political consciousness. (p. 123; emphasis in original)

Related to the themes of heterogeneity and historical contingency is an appreciation of how groups may engage in strategic essentialism (Fuss, 1989). Strategic essentialism refers to when subordinated groups employ biologically or culturally based arguments from an oppositional position. This can be seen, for example, by groups reveling in a newly re-created past identity, which is usually an effort to resist changing conditions. Thus, the tendency to cultivate "back to the roots" identities and movements is a reflection of the present informed by the past (Moghadam, 1994, p. 12; Rutherford, 1990). Although the development of a people of color identity by the environmental justice movement is a response to changing political landscapes, how does one interpret the resistance to this new identity on the part of rank-and-file movement members? Their hesitancy to adopt a new, seemingly useful political identity should not be construed as a "narcissistic cele-bration of culture and identity" (Gilroy, 1992, p. 56) but rather may emanate from activists' belief that it is not in their best interest to do so. Because people now recognize the diversity and changing nature of the many identities contained within each of us, it follows that there may be resistance to certain forms of identification.

Sparse attention has been directed to understanding how people may resist the adoption of particular identities or the ambiguity and confusion surrounding that process. Brah (1992) has dealt somewhat with this question in her examination of how Asians in the United Kingdom have problematized the category *black*. Once used to denote all non-whites to present a unified image and to avoid the reifying of phenotypic variation, the term has also served to deny an Asian cultural identity. In Los Angeles, by contrast, various groups are hesitant to move beyond their traditional ethnic identities because they fear that their uniqueness will be lost, in either history, needs, or political agenda. For example, Gray (1993) has argued that most blacks are a long way from acknowledging their numerous identities simply because their material reality is under attack from all quarters. Moreover, Uhlaner's (1991) survey of various Southern California minority groups found that African Americans, Mexican Americans, and Asian Americans all had different perceptions of how much discrimination they encountered and its consequences. These findings suggest that the experience of racism may be too fragmented to serve as the basis of a larger identity and collective action.

To a limited degree, various constituencies experience differing forms of oppression that militate against the formation of a collective identity and action. For example, one activist described the need for "self-organizing," that is, organizing solely among one's own ethnic group, and particularly among the most marginalized.

> The self-organization of African Americans is a legitimate form of organizing. I am not a nationalist. So I don't see the solution to our problems as the separation of races. But I do believe, whether it is African Americans, Latinos, or women, that self-organizations are a necessary step towards building a movement.

Nevertheless, he did not see this activity as precluding the building of a multiracial movement.

> I think there are stages of development in terms of building a multiracial movement and building multiracial organizations, and I think the self-organization of oppressed people is a legitimate state in that process.

This appears to be true among all groups, but for slightly different reasons, depending on the specific nature of their oppression. In the case of Latinos, language is the primary form of marginalization, supplemented by anti-immigrant fervor and discrimination against nonwhite bodies. Simply put, monolingual Spanish speakers are largely excluded from all public participation; agency outreach is awkward at best. Consequently, *las Madres* (who include monolingual Spanish, English, and bilingual members) fill a necessary niche. MELA's water conservation project was in fact initiated because of the Metropolitan Water District's inability to connect with large portions of the Latino population.

African Americans' unique concerns stem not only from the extreme hardship facing segments of the African American population but also from historically entrenched racism and competing claims from other racial/ethnic groups.

> Sometimes when we talk about our commonality as people of color and our common oppres-
> sion, we lose sight of the specific forms of oppression and what has been done to us historically.
> And I really believe that there is something deep that has happened to us Black people in this
> country as a result of our specific oppression. . . . There is a lot of that (Black self-organizing)
> in terms of "no one understands us like we understand ourselves." Or, "Every time we try to
> get something we're quickly merged into half a dozen other ethnic groups. We're tired of that."
> Yes, there is some of that.

Thus, although many groups suffer from racism and oppression, the lived reality for many individuals is far more complex. Why and how should African Americans unite with a group that is prejudiced against them and diverts necessary resources? Why should Latinos collaborate with a group that supports anti-immigrant legislation? Conventional wisdom holds that both oppressed groups would recognize either whites and/or capital as responsible for their marginalization, but such assumptions disregard the real material and political differences of each group's circumstances. This brings us back to Gray's (1993) observation that African Americans are more than ever choosing to assert a unified, essentialist black identity, *largely because they feel it is in their best interest to do so.* Although subordinated groups possess various and shifting identities, they may resist new ones, particularly those that they perceive might dilute or jeopardize traditional identities.

Ironically, although it may be difficult for low-income African Americans and Latinos to adopt new identities and forms of political action, this is not the case for the minority middle class and professionals, who have readily embraced the identity offered by "people of color" and the multiracial agenda that it implies. There has been a proliferation of organizations and activities surrounding environmental justice issues—but geared to the needs, aspirations, and agendas of middle-class/professional people of color (Environ-mental Careers Organization, 1992). The identities these organizations adopt are apparent from the names of the groups, including the Ethnic Coalition and the Ethnic Advisory Council of the SCAQMD. These groups do not exist as political organizations or movements but simply aim to ensure that people of color are informed about local environmental issues and have the opportunity to voice their concerns to decision makers in the hope of achieving a "negotiated consensus." The impetus for the formation of the Ethnic Coalition is instructive. During the 1980s, critical urban environmental issues were being debated in Los Angeles, including air quality, regional transportation, and local job-housing balance. Several nonwhites who worked with developers, public utilities, and government realized that long-term regional decisions were being made almost entirely by Anglos. Thus, with backing from the Gas Company and institutional support from such entities as the University of Southern California, the Ethnic Coalition was formed, according to one of its cofounders, because the problem with Los Angeles, as multiethnic as it is, the key decision makers are still in the Anglo community. . . . Our approach is that whatever the issue is that's on the table, ethnic people need to be involved. . . . We have to as ethnic groups . . . pull our power in order to be effective. It's not just about you or me knocking on doors. I mean we're going to have to be *coordinated.*

Clearly, there are political differences between those who advocate negotiated consensus and those who demand the cessation of pollution or hazardous land uses.

The pursuit of an antiracist agenda by groups of different races requires an identity shift. One corporate-based activist relayed how her identity changed from "Black" to a "person of color": "My thing is that when they look out there and they discriminate against minorities, it's not necessarily just because I'm Black, but it's just all of us minorities out there." This makes perfect sense because the African American, Asian American, and Latino middle classes are far more secure than their socially and spatially marginalized counterparts. The more affluent have the luxury of taking the risk of developing an alternative and new identity, one not always geared to the immediate needs of the ethnic group into which one was born.

Both the grassroots and professional activists are employing different but unitary representations of people of color, or particular ethnic groups, with scarce consideration to class differences. Thus, the Watchdog believes it speaks as much for people of color as the SCAQMD Ethnic Advisory Council. Both discourses are largely developed in opposition to racism, but with different objectives. One grassroots activist believed that this lack of class analysis and identification inhibited the formation of multiracial identities and political action:

I think that the thing that would help bridge races together in common work is an understanding of common histories and oppression, and a lot of it is having a class analysis which is sorely lacking in this society. . . . So the ruling class has set up a situation where we have a huge middle class which has an interest in maintaining the status quo, even though they are not where they want to be. . . . They see themselves as middle class, not working class, there's a big difference. . . . So everyone else is scrambling to hold on to their little piece, but with no discussion about class. It just boils down to, "those people are moving into my neighborhood, there were three of them last month and now there are six, they're somehow a threat to me." That's where the debate is right now.

This is an important point in understanding larger attempts at political mobilization and the future of environmental justice in Los Angeles. Because the people-of-color middle class is able to clearly articulate a united and coherent agenda, it may ultimately wield more power than grassroots organizations that (although they have sharp analyses of their individual situations) are forging a united opposition only to a limited degree.

On the basis of this analysis, it should be apparent that the conventional wisdom routinely espoused by antiracist and grassroots activists concerning the formation of a broad-based social movement must be reconsidered. The notion that if only linkages were properly articulated, then various oppressed groups would find common ground is weak at best and denies the power associated with different forms of oppression. The different forms of racism experienced by various groups are at least as significant as both groups experiencing racism and economic subordination. Moreover, the use of local culture, such as language, needs to be reexamined in light of postmodernity when there is no longer a firm line between the oppressed and the oppressor. Although the appropriation and use of marginalized culture will always be a tool of resistance, the politics surrounding its use are becoming increasingly murky and problematic. Given the reality of multiple oppressed groups occupying various positions, the emphasis on difference as a strategy must be reconsidered in a new light.

Conclusion

Nonwhites' participation in the environmental justice movement has coalesced around an identity as people of color. Los Angeles has been a site of much of this activism and is often considered a leader both in environmental justice battles and in forging a multiracial movement. Although many leaders have a strong commitment to building a genuinely diverse and multiracial environmental justice movement, the reality is that the discourse developed by the leaders is not reinforced by a substantive multiracial identity and activism by the larger membership. The resistance to even a secondary identification as people of color exists for a number of reasons, including language diversity, intergroup prejudice, the perceived need for self-organizing, and a lack of class analysis. Although grassroots activists remain important political players and their organizations have continued crucial work in their respective communities, their inability to pursue a new identity has contributed to their limited efficacy in articulating the needs and concerns of marginalized communities on citywide and regional environmental issues. Instead, this role has been partially usurped by professional minority activists who often do not live in the affected communities and espouse more conservative politics but yet are seen as legitimate representatives of the communities in question. One possible reason why working-class nonwhites cling so tenaciously to traditional racial/ethnic identities is because of the increasingly difficult material circumstances of life, particularly for persons who are poor. This calls into question the purportedly multiracial nature of the movement. Thus, although minority environmental justice activists have made a tremendous impact in Los Angeles, the future is uncertain.

Although this chapter has focused on environmental justice activism, it is quite possible that the political dynamics outlined above will be found in other arenas—anywhere identity politics are emphasized. Moreover, it appears that given the increasingly fragmented and diverse nature of Angelenos, identity politics will continue to be a powerful force in the local scene as various groups organize along specific lines and make particular claims. Although there is no denying that identity politics can be a powerful and useful tool to incorporate historically marginalized groups into the dominant society, we should also be aware of whose interests are really being represented and how that representation is occurring.

Notes

1. Smog is a photochemical oxidant resulting from a reaction between carbon monoxide, ozone, nitrogen dioxide, and particulate matter. Automobiles produce a majority of the constituent elements of smog (59% of nitrogen dioxide, 87% of carbon monoxide, and less than 6% of particulate matter), but industry also contributes a significant amount. Air toxins, in contrast, are overwhelmingly produced by industrial facilities. Although smog and air toxins fall under a different set of regulatory categories, in the eyes of the general population it is, of course, all air pollution.

2. The criteria pollutants include lead, sulfur dioxide, nitrogen dioxide, carbon monoxide, ozone, and particulate matter. The plan was intended to bring the region into federal compliance for carbon monoxide, sulfur dioxides (not a significant problem in Southern California), nitrogen dioxide, and lead but not for ozone and particulate matter. The latter two are so severe that far more time and stringent measures are needed to bring the region into compliance.

3. Because of the polarized nature of the Southern California economy, minority communities *are* the low-income communities. Although there are poor Anglos in Los Angeles, they do not constitute a large percentage of the total poor population, nor are they spatially concentrated (Ong & Blumenberg, 1992).

4. This does not mean that racial/ethnic minorities were not involved in environmental issues but that they were not part of the larger, mainstream environmental movement. For examples of such activism, see Pulido (1996).

5. According to the U.S. Bureau of the Census (1990), Los Angeles County was approximately 40% Anglo, 40% Latino, 11% African American, and 10% Asian-Pacific Islander.

6. My analysis is limited to Latinos and African Americans for several reasons. First, as the two poorest groups in Los Angeles, they are marginalized by both racism and economic subordination. Second, although some scholars have focused on African American-Latino competition (Johnson & Oliver, 1989), many progressives hope that they will unite and effectively challenge local social relations. Finally, both groups have been quite active in environmental justice issues nationally. Asians are not included because although Asian poverty is a very real problem, Asians are much more economically diverse and often engaged in antagonistic class relations with African Americans and Latinos, such as the "ethnic middleman" (see Light & Bonacich, 1991), thus making the possibility of interethnic activism far more difficult. Moreover, Asians have not participated significantly in the environmental justice movement. Asian participation is much more marked among the professional/middle class than among low-income activists.

7. Los Angeles City Energy Recovery Project.

8. Although Texaco emits 256,576 pounds of air toxins annually, it is not the largest emitter in the region (U.S. Environmental Protection Agency, 1994).

9. Citizens for a Better Environment recently changed its name to Communities for a Better Environment in an effort not to exclude noncitizens.

References

Austin, R., & Schill, M. (1991). Black, brown, poor and poisoned: Minority grassroots environmentalism and the quest for eco-justice. *Kansas Journal of Law and Public Policy, 1,* 69-82.

Bloch, R., & Keil, R. (1991). Planning for a fragrant future: Air pollution control, restructuring and popular alternatives in Los Angeles. *Capitalism, Nature & Socialism, 2,* 44-65.

Brah, A. (1992). Difference, diversity and differentiation. In J. Donald & A. Rattansi (Eds.), *"Race," culture and difference* (pp. 126-145). Newbury Park, CA: Sage.

Brown, P., & Masterson-Allen, S. (1994). The toxic waste movement: A new type of activism. *Society and Natural Resources, 7,* 269-287.

Bullard, R. (1990). *Dumping in Dixie.* Boulder, CO: Westview.

Burke, L. (1993). *Environmental equity in Los Angeles.* Unpublished master's thesis, University of California, Santa Barbara.

Capek, S. (1993). The "environmental justice" frame: A conceptual discussion and an application. *Social Problems, 40,* 5-24.

Citizens for a Better Environment. (1989). *Richmond at risk: Community demographics and toxic hazards from industrial polluters.* San Francisco: Author.

Environmental Careers Organization. (1992). *Beyond the green: Redefining and diversifying the environmental movement.* Boston: Author.

FitzSimmons, M., & Gottlieb, R. (1993). Environmental planning and policy in the Los Angeles region: Openings and opportunities. In A. J. Scott (Ed.), *Policy options for Southern California* (pp. 63-83). Los Angeles: University of California, Lewis Center for Regional Policy.

Fuss, D. (1989). *Strategic essentialism.* New York: Routledge.

Gilroy, P. (1992). The end of anti-racism. In J. Donald & A. Rattansi (Eds.), *"Race," culture and difference* (pp. 49-61). Newbury Park, CA: Sage.

Gilroy, P. (1993). It ain't where you're from, its where you're at. In P. Gilroy (Ed.), *Small acts* (pp. 120-145). London: Serpent's Tail.

Gray, H. (1993). African American political desire and the seductions of contemporary cultural politics. *Cultural Studies, 7,* 364-373.

Gregory, S. (1993). Race, rubbish, and resistance: Empowering difference in community politics. *Cultural Anthropology, 8,* 24-48.

Hofrichter, R. (Ed.). (1993). *Toxic struggles.* Philadelphia: New Society.

Jackson, P. (Ed.). (1987). *"Race" and racism.* London: Allen & Unwin.

Johnson, J., & Oliver, M. (1989). Interethnic minority conflict in urban America: The effects of economic and social dislocations. *Urban Geography, 10,* 449-463.

Kay, J. (1994). California's endangered communities of color. In R. Bullard (Ed.), *Unequal protection* (pp. 155-188). San Francisco: Sierra Club.

Kerr, M., & Lee, C. (1993). From conquistadors to coalitions. *Southern Exposure, 21,* 8-19.

Labor/Community Strategy Center (LCSC). (1989). *L.A. fights for breath.* Los Angeles: Author.

Lee, C. (1992). Toxic waste and race. In B. Bryant & P. Mohai (Eds.), *Race and the incidence of environmental hazards* (pp. 10-27). Boulder, CO: Westview.

Lents, J., & Kelly, L. (1993, October). Clearing the air in Los Angeles. *Scientific American, 269,* 32-39.

Light, I., & Bonacich, E. (1991). *Immigrant entrepreneurs: Koreans in Los Angeles, 1965-1982.* Berkeley: University of California Press.

Mann, E. (1991). *L.A.'s lethal air.* Los Angeles: Labor/Community Strategy Center.

McGurty, E. (1995). *The construction of environmental justice: Warren County, North Carolina.* Unpublished doctoral dissertation, University of Illinois, Urbana.

Moghadam, V. (1994). Introduction. In V. Moghadam (Ed.), *Identity politics & women* (pp. 3-26). Boulder, CO: Westview.

Nash, J. (1992). Interpreting social movements: Bolivian resistance to economic conditions imposed by the International Monetary Fund. *American Ethnologist, 19,* 275-293.

Ong, P., & Blumenberg, E. (1992). *Income and racial inequality in Los Angeles.* Unpublished manuscript, University of California, Urban Planning, Los Angeles.

Pardo, M. (1990). *Identity and resistance: Mexican American women and grassroots activism in two Los Angeles communities.* Unpublished doctoral dissertation, University of California, Los Angeles.

Pulido, L. (1996). *Environmentalism and economic justice: Two Chicano struggles in the Southwest.* Tucson: University of Arizona Press.

Russell, D. (1989, Spring). Environmental racism. *Amicus Journal,* 22-32.

Rutherford, J. (1990). A place called home: Identity and the cultural politics of difference. In J. Rutherford (Ed.), *Identity, community, culture, difference* (pp. 9-27). London: Lawrence & Wishart.

South Coast Air Quality Management District (SCAQMD). (1989). *Summary of 1989 air quality management plan.* Diamond Bar, CA: Author.

South Coast Air Quality Management District (SCAQMD). (1992, August 7). *Report to the governing board regarding the recommendations of the special commission on air quality and the economy and recommendation to adopt certain recommendations* (Agenda #35). Diamond Bar, CA: Author.

Szasz, A. (1994). *EcoPopulism: Toxic waste and the movement for environmental justice.* Minneapolis: University of Minnesota.

Taylor, D. (1992). Can the environmental movement attract and maintain the support of minorities? In B. Bryant & P. Mohai (Eds.), *Race and the incidence of environmental hazards* (pp. 28-54). Boulder, CO: Westview.

Uhlaner, C. (1991). Perceived discrimination and prejudice and the coalition prospects of Blacks, Latinos, and Asian Americans. In B. Jackson & M. Preston (Eds.), *Ethnic and racial politics in California* (pp. 339-371). Berkeley: University of California, Institute for Government Studies.

United Church of Christ. (1987). *Toxic waste and race in the United States.* New York: United Church of Christ, Commission on Racial Justice.

U.S. Bureau of the Census. (1990). *1990 Census of population and housing* (Summary Tape File 3). Washington, DC: Author.

U.S. Environmental Protection Agency. (1994). *1987-1992 toxic release inventory* (EPA 749/C-94-001). Washington, DC: Author.

U.S. General Accounting Office. (1983). *Siting of hazardous landfills and their correlation with racial and economic status of surrounding communities.* Washington, DC: Author.

Environment, Economy, Equity

LILLIAN KAWASAKI
City of Los Angeles Environmental Affairs Department

In dealing with the environment today, we have to address and integrate what I call the three E's: Environment, Economy, and Equity. On the last *E,* a livable Los Angeles must be inclusive of ethnically and culturally diverse communities. Our environmental agenda must be built on an environmental justice-equity principle based on who pays and who benefits. Much more work needs to be done.

Radical Reform Versus Professional Reform in American Schools

A View From Southern California

GUILBERT C. HENTSCHKE

Overview

The rising tide of dismal education reports is by now familiar to most Americans. American students are falling behind. They are doing worse today than students have done historically; our children are performing much worse than children from nations with whom they compete internationally. . . . Perhaps most dispiriting of all is the attempt by some educators to claim that there is nothing wrong with American education, but [that] there is something wrong with the dismal reports about it. (Gerstner, Semerad, Doyle, & Johnston, 1994, p. 8)

During the past year, ECS [the Education Commission of the States] examined the state's education system from a variety of angles and perspectives. . . . The picture that emerged, in many ways, is heartening. Awash in new demands and operating under many financial constraints, California's school system nevertheless not only maintained but also improved its performance and efficiency over the past decade. (Education Commission of the States, 1995, p. 5)

In California, a majority of elementary and high school students recently failed for the second year in a row to demonstrate minimal reading, writing and math skills in objective tests. The state's response was to scrap the tests as too difficult and convene a task force. (Flanigan, 1995, p. D1)

> Contrary to what [education] reform[er]s have been claiming, the central failure of our
> education system is not inadequacy but excess: Our economy is being crippled by too much
> spending on too much schooling. [Educational] reform is a hoax. . . . It is [also] outrageously
> expensive. (Perelman, 1992, p. 24)

Is there an education problem? If so, what is the problem? Given that problem, can it
be solved? Should it be solved? If so, by whom? How do the educational circumstances
in Southern California, especially Los Angeles, inform these education dilemmas? Despite the presence of multiple contradictory analyses and widely varying proposed
remedies for public schooling, only two broad agendas exist, and, given the emergence
of three trends and circumstances, one policy option emerges.

This assertion constitutes three interlocking objectives in this chapter: (a) to describe
the two competing reform agendas for public schools, (b) to identify three circumstances
that are influencing the outcome of the competition between these two school reform
agendas, and (c) to seek to identify the central policy issue that is emerging in Los Angeles,
throughout California, and across the country. Amid the growing cacophony about efforts
to improve the educational achievement of American children during the last two decades,
two broad and opposing streams of school improvement rhetoric dominate: a large and
visible reform agenda supported through a loose confederation of largely education
professionals and a smaller but more radical reform agenda supported and advanced
through a loose confederation of individuals from a variety of sectors other than education.
"Professional reform" (by definition) describes the efforts of the education profession to
improve its professional practices, but the agenda of this group is increasingly challenged
by "radical reformers" (composed largely of individuals—parents, politicians, and businesspeople—who are not in leadership positions in the education profession) who view
the professional reform agenda as inadequate at best and, at worst, as much a part of the
"public school problem" as the public school system itself.

For radical reform to succeed, professional reform will have to fail. Although professional reform has so far continued to dominate efforts to improve schools, its continued
dominance is threatened by the interaction of three circumstances: (a) the perception by
local communities that local schools are not responsive to local aspirations for them; (b)
increasing differentiation among communities, signaling growing variabilities in their
cultures, ethnic composition, and socioeconomic status; and (c) strategies for professional
school reform that do little to address the growing lack of connection between local
schools and local communities. Extreme forms of these circumstances exist in Southern
California, where communities no longer vote on annual school budgets, where variations
in culture, class, and income levels are among the highest in the country, and where as
many professional reform initiatives have been launched as anywhere else in the country.

In pursuing the worthy goal of treating all children in society equally, regardless of
socioeconomic background, the public school financing systems have reduced, and, in the
case of California, all but eliminated, the most fundamental accountability mechanism
between schools and their communities—the local vote on the school budget. As the
socioeconomic, cultural, and demographic makeup of communities continues to fragment,

connections between communities and largely middle-class Anglo educators in their public schools will further erode. Finally, professional school reform efforts will not help much because they continue to be designed and funded by agents external to local school and civic communities, for example, by foundations as well as state and federal government agencies. The resulting agenda for school reform in a local community is largely one agreed on between professional educators and those organizations. Some parents and community partisans, who may already feel incapable of influencing local school policy, are not significantly won over by professional reform efforts.

Southern California has served as a stage for playing out the competition between the two school reform agendas. In light of changing circumstances here, the reform agenda advocated by professional educators is increasingly threatened by radical school reform. Parents and local civic and political leaders are seeking greater influence over local school policies and practices than current professional school reform efforts promise them. By instituting a form of local voice over local school spending and, hence, over local school policies and practices, citizens, including professional educators, can reestablish a continuing dialogue at the local community level about what they want their schools to accomplish for them. Given the growing differentiation among local communities, especially in the greater Los Angeles area but also throughout the state, it would serve us well to localize decisions on public school spending and operation as an alternative to privatization.

Two Parallel Streams of School Reform

Two streams of school reform have been flowing parallel to each other for some time. Both base their agenda on a critique of the current system of American public education, but the critiques and reform platforms of each stream differ sharply from each other. The first stream of school reform is highly visible in print, on podiums, and in daily conversations among educational leaders. It is revealed in the work of commissions, task forces, and offices at the federal, state, regional, and school district levels. The language of this reform stream is familiar but evolves through time. The philosophical anchor of professional school reform has moved from one of schooling processes to one of student performance. This anchor is loosely connected to similar discussions about the changing nature of work in other organizations. The revolution in workplace organization and increased worker responsibility has included educational institutions as places of adult work and caused educators to evaluate and redesign the nature of work in schools.

The fundamental argument of professional reform has three interconnected parts: (a) Schools have been run on the basis of specified processes; (b) student achievement under this system is not at levels that are acceptable today; and (c) schools should now be organized and run in such a way as to place a premium on student performance, no longer on the basis of compliance with procedural requirements. The argument for the historical system was compelling. Schools were historically organized on an "industrial model" wherein scarce expertise was located at the top of the organization, and detailed directions were passed down the organization chain of command to the line worker (teacher), who carried out the required procedures of instruction. Through time, these directions became codified as a set of procedures, that is, production processes that guided the work of educators. Instruction was delivered in classes of a certain size, with a specified set of texts, covering a specified curriculum context, during a specified number of days, and so on. By adhering to the stipulated process of schooling, the educational system was deemed to be educating children and youth. Educators, although hardworking, were not required to assume responsibility for the effects of their instruction; instead, they were responsible for following procedures designed by superordinate boards and governments. At the risk of extreme generalization, one might say that the traditional system of schooling was built around three components: The "best" people in the profession rose to the top of the system, they ensured that those below them complied with rules of behavior designed to deliver instruction, and all of the incentives (rewards and penalties) in the organization were designed to ensure compliance with procedures.

The shift from procedures to performance in effect turns the traditional organization on its head. Instead of procedures (how the educator is supposed to behave), the focus on the system is on performance (what the student is supposed to know and be able to do). Relatedly, instead of the "best" people gravitating to the top of the organization, the new public school organization would be restructured so that the "best" people are those in direct contact with children. All the incentives of the organization then have to be realigned so that those educators closest to the children (frontline workers) will do everything in their power to bring children up to those (high) levels of student performance. One of the most comprehensive treatises on professional reform is Marshall and Tucker's (1992) *Thinking for a Living: Education and the Wealth of Nations.* Currently, the text of this reform stream is laden with additional words and phrases such as *systemic, break the mold, authentic,* and *performance-based,* all of which with others have been layered on top of slightly older words and phrases such as *site-based, cognitively complex,* and *teacher-empowered.*

Issues addressed by these adjectives in this stream of school reform include but are not limited to developing uniformly high student achievement standards for all students except those who are the most severely disabled; installing student testing programs that incorporate greater proportions of student work in "student portfolios" and lesser proportions of mark-sense, multiple-choice, and fill-in-the-blank test forms; driving this high performance via much more challenging curricula; greatly enhancing the authority and accountability of educators at the school site to encourage them to undertake these challenges; engaging parents in the education of their children; linking primary health, mental health, and social services programs much more intimately to the children and families served by schools; creating clearly articulated programs between school and the workplace as well as between school and college; and significantly increasing professional development (educator job-related training) to bring about all of these changes.

Examples of professional school reform involving schools in Southern California are numerous. Among state initiatives are two authored by then State Senator Gary Hart: (a)

Senate Bill (S.B.) 1274 (1989), Demonstration of Restructuring in Public Education, a multiyear competitive grant program designed to bring about "whole school" reform of the type described above in about 100 schools around the state; and (b) charter school legislation, (S.B. 1448, 1991) which requires school district-level approval, enabling up to 100 school communities a degree of increased autonomy to pursue agendas not dissimilar from the agenda described above. These and other state legislative initiatives have been reinforced by a steady and overlapping stream of policy documents including *The Unfinished Journey: Restructuring Schools in a Diverse Society* (Olsen et al., 1994); *Rebuilding Education in the Golden State: A Plan for California's Schools* (Policy Analysis for California Education, 1995); as well as *Rising to the Challenge: A New Agenda for California Schools and Communities* (Education Commission of the States, 1995).

In addition to state-initiated professional school reform, Southern California has also involved itself in national and even local variations. Examples of national initiatives (both public/governmental and private/philanthropic) that have connected to all or parts of Southern California include the Coalition of Essential Schools, Accelerated Schools, the New American Schools Development Corporation,[1] the Urban Systemic Initiative (Los Angeles Unified School District [LAUSD]),[2] and the Annenberg Challenge (Los Angeles County).[3] Local examples of single-district initiatives that incorporate roughly the same set of seven educational reform principles include Los Angeles Educational Alliance for Restructuring Now (LEARN), which targets the LAUSD (Pyle, 1995), and New Directions, a similar reform effort on behalf of neighboring Pasadena public schools (Pasadena Unified School District, 1993).

Despite their complexity, variability, and continually evolving nature, these and other similar "systemic" reforms share a few general characteristics: (a) They incorporate a list of changes in educational practice roughly similar to the list described earlier; and (b) they are endorsed by the majority of educational leaders and the civic, business, and legislative officials with whom they interact on a regular basis. In fact, many of these professional school reform initiatives peacefully coexist within districts and even within individual schools, testifying to their compatibility, if not their sameness. This reform stream is "professional reform" in that it has been formulated by professional educators, that is, those individuals who have extensive experience working in and on behalf of the current system of public schooling. In other words, there exists a rough but nonetheless identifiable consensus among leaders in the education system about the shape of reform of public education.

This is not to say that there are not occasional skirmishes within various parts of the professional school reform camp, such as concern by school district bargaining units in the LAUSD about representation in charter schools and concern by central administration in the LAUSD about the degree of autonomy that should be permitted to schools that have joined the district's reform program. Rather, the point argued here is that there is a reasonably broad consensus within the leadership of the profession on the grand principles staked out earlier. Professional reform is a collection of principles, assumptions, and associated actions that is sufficiently acceptable to and supported by enough of the leaders associated with the education profession. It constitutes the dominant agenda for public school change, having been formulated by those close to and in the current system of public education.

The second, smaller stream of school reform differs from the first in that it is generally not sanctioned by the majority of these same individuals and groups—call it the "radical" stream of school reform for purposes of delineation. It, like the first stream, has its own critique of the current system and its own language, issues, and projects. Chief among the phrases associated with the radical reform stream are *parent-empowered, market-driven, antimonopolistic, choice-based,* and *privatized.* The primary presumption of this smaller, less publicly articulate group is that public schools are inherently incapable of reform from within. Any changes recommended by educators serve the interests of educators at the expense of children and families. Only changes that make the education establishment more accountable to taxpayers and parents are worthy of consideration. High among the list of preferred changes are those that foster various forms of competition through choice, both for parents in selection of schools and for districts and teachers through contracting provisions.

Specific issues of the radical reform group include contracting out for instructional services, creating and implementing voucher initiatives, abolishing teacher tenure, creating private, company-based schools for children of employees (Hadley, 1995, p. 4), and breaking up the LAUSD, the second largest district in the United States, composed of about 650,000 pupils housed in about 650 schools. Projects include lobbying for legislation to bring about these changes as well as mounting initiative campaigns to place measures on state ballots. These efforts are led by, among others, individual business executives, conservative educational policy organizations, conservative politicians, and parent groups that are unaffiliated with formal long-established parent-teacher organizations. All these radical reform efforts are largely opposed by a majority of the leadership of educational institutions, especially those associations that represent teachers, school board members, school administrators, and other school employees. From the perspective of these and other professional groups, these remedies are not simply too radical, they are wrongheaded.[4]

The radical reform stream has been running parallel to the much broader, professional reform stream for at least a decade. These streams are parallel in that supporters of one stream tend to be opponents of the other. Individuals concerned with K-12 education reform find themselves largely in one stream of reform or the other, but not both.[5] There are, of course, some who anguish over such a choice. This chapter is directed in part at the evolving conditions that will influence their choices one way or the other. These two reform streams are also parallel in the sense that both have existed and evolved during roughly the same period. Between 1988 and 1995, professional reform efforts evolved alongside radical reform efforts in the LAUSD.

During this relatively short time, a variety of professional reform initiatives were created, operated, "scaled up," and (in some cases) passed into history: Shared Decision

Making was followed by School-Based Management, which was supplemented by S.B. 1274 (1989), followed by LEARN, followed by and complemented by Los Angeles Learning Centers (one of the nine proposals awarded nationwide by the New American School Development Corporation), and complemented even further by the Los Angeles Metropolitan Project (one of the urban K-12 initiatives funded by the Annenberg Foundation) and the Los Angeles Systemic Initiative (one of the reform initiatives of the National Science Foundation).

During those same 8 years, a separate, radical reform agenda was pursued by a different set of people (largely but not entirely noneducational leaders) through a wholly different set of initiatives. Proposition 174 was placed on the California ballot in 1993 to provide vouchers to school-age children. This was followed by two attempts to break up the school district (the second is still under way) and later by a pledge by Governor Pete Wilson to do away with teacher tenure. Currently, several additional voucher proposals are being prepared for the 1996 ballot that attempt to remedy the perceived weaknesses of the earlier voucher initiative.

Although the two reform streams are distinguishable, indeed distinct from each other, they share some elementary common characteristics, and furthermore, they are at some level dependent on each other. How are they similar? Both reform groups argue for extreme improvement, albeit with critiques of varying degrees of harshness. The mildest criticism from both camps is that "public schools have not deteriorated, but also they have not changed in response to changes in society." As critics get more specific, they identify problems in ways that suggest the solutions specific to one or the other reform agenda. Both reform camps criticize, for example, the "command and control" model of administration that characterizes public schooling. But although the professional reform group sees site-based management as a remedy, the radical reform group sees market discipline via vouchers as a remedy.

The arguments marshaled to support each reform stream are both high-minded, for example, "to benefit children." At the same time, the pro arguments of both professional and radical reformers at some levels promote the self-interests of parts of their reform constituencies, for example, advancing the agenda of teachers unions (professional reform) through greater authority for teachers and silence on teacher tenure and on pay for performance *versus* supporting parents with current or future children in public and/or private schools of choice (radical reform) through voucher mechanisms. This is also true of the con arguments marshaled by each reform to criticize the other. The professional school reform stream views the radical stream as a "threat to the very concept of public education as we know it." The radical school reform stream views the professional school reform stream as "protecting monopolistic bureaucracies."

Despite the vitriol between the two camps as well as the nonoverlapping nature of the action agendas and participants in each, an ironic mutual dependence exists between the two reform camps: The energy and actions of each reform camp are influenced by the energy and actions of the other. At any point in the battle of the two reform groups, proponents of radical education reform are strengthened to the degree that the professional reform agenda is seen as ineffectual. Proponents of professional school reforms, on the other hand, acknowledge that they make much greater progress on their agenda when tangibly "threatened" by impending radical reform initiatives. California Secretary of Education and Child Development Maureen DiMarco warned that professional reform agendas in the state have little time to take effect before "far more Draconian measures—a voucher, for example—are put before a public that is sick and tired of rhetoric and fruitless disputes about the schools."[6] The reverse is also true: "The commitment to significant structural reform clearly is on the wane in the wake of the lopsided defeat of [the school voucher initiative] as well as the rejection of legislation—at least so far—to break up the LAUSD" ("Reform Roadblock," 1995, p. 12). Radical school reform can succeed only when/if professional school reform sufficiently fails; professional school reform can only succeed when/if sufficiently threatened by radical school reform.

Initiatives to break up the LAUSD (radical reform) have been designed in part to make the district more responsive to localities, and, on occasion, the initiatives have been perceived as a real possibility (real threat) by education professionals in the district. District educators have been able to blunt the arguments of the pro-breakup group by initiating their own form of restructuring in the district—organizing the entire district into 27 "clusters"—each composed of at least one high school along with the middle and elementary schools whose students tend to graduate largely into that high school. In their own version of "breaking up" the district, LAUSD officials were able to describe the benefits of increased local sensitivity through the cluster leaders, blunting the arguments for a legislated breakup of the district. The threat of breaking up the district has been blunted only temporarily, however, because local parent and community groups discern a major difference between an administrative reorganization of the district and the creation of numerous, smaller districts, each with its own board of education, its own budget, its own problems, and its own school reform agenda.

The battle to improve schools, in this view, is not so much between proponents of no change in public schools and reform-minded leaders in the education profession. Rather, it is between a brand of school reform that public school educational leaders actively and publicly support and a brand of school reform that the majority of these same individuals actively resist (or at least feel they must publicly resist). The battle is not with the public at large; it is between different brands of reform, with each side seeking to win over a sufficient plurality of the public. If too little progress is made via professional school reform efforts, greater public support will be provided to radical school reform efforts.

Educators state simply and routinely when discussing reform in Southern California, "If the schools won't reform themselves from within, then they will be reformed by outsiders." In this view, professional reform will, by definition, be the agenda of public school leaders unless and until their arguments and behaviors are so "challengeable" that radical reformers are able to make a successful assault. As long as professional reform moves fast enough in the "right" direction, radical reform will not succeed in its occasional attacks. On the other hand, to the extent that improvement in schooling (read "improvement in publicly reported test scores") is perceived as too slow, radical reform agendas gain in popularity.

Issues Challenging Professional School
Reform in Southern California

Despite periodic challenges from radical reformers, professional reformers continue their hegemony over radical reformers. Interacting circumstances in Southern California, however, make it increasingly difficult for professional reformers to retain this advantage over radical reformers: rapidly growing numbers and proportions of poor children who require greater support from the public school system while at the same time contributing to greater heterogeneity in local communities; a steadily declining fiscal support base for schooling and for services that act to support success in schooling coupled with a fundamental fiscal disconnection between schools and their local communities; and traditional mechanisms for improving schooling that are necessarily limited in scope and contribute little to connect schools and local communities.

These circumstances work against the reform agenda of educational professionals by challenging several implicit principles of professional reformers: All students and all school communities should be treated identically, which requires that equal amounts of financial inputs be made available to all similar students; public financial support of public schooling should derive only from publicly raised tax dollars expended through publicly operated institutions, but local communities should not be able to raise more tax dollars for their schools; and the system of public schooling should be improved through mechanisms that start with the most willing and capable schools and districts and then spread to the others.

How do the forces work against the principles? With greater proportions of children coming from poverty and "at-risk" backgrounds, nonschooling factors that have always contributed to school success are less in evidence. Equal per pupil expenditures do not significantly and directly influence those factors that negatively affect the likelihood of success of poor kids in school. Relatedly, when the only vehicle for supporting public education is publicly raised tax dollars, and when those dollars are declining relative to other regions and there appears to be no way to reverse that trend, there will be increasing willingness to search for other resources to support public schooling. This includes both local tax levies and heretofore private and voluntary sources of support such as from anxious, concerned parents with discretionary income. Finally, the professional school reform strategy of improving the achievement levels of all students by first providing incremental grant support to a small fraction of the most willing and capable among the public schools, and then hoping to somehow spread success to other schools, may simply not be sufficiently swift or effective in the future. This is especially true if, as radical reformers argue, no fundamental changes in rights have been made, for example, in teacher tenure and parent choice.

The strength of these forces and their corresponding impact on professional reform principles are, arguably, greater in California, especially Southern California, than in other parts of the country. The scale of each is described next.

Equal Treatment of Similar Students
Versus the Growing Proportion of Needy Students

"The accident of residential location should not determine the educational resources that the state makes available to children." This legal principle, derived from the 14th Amendment, has been extensively applied to justify school finance formulas that mitigate the school funding effects for students living in relatively poor school districts. On top of this financial and legal foundation, modest additional resources have traditionally been targeted by the federal government to a relatively small fraction of students whose personal (and family) circumstances required additional support.[7] How would this principle of equity evolve if there were suddenly many more needy children to serve and, in addition, the proportion of needy students were to grow rapidly until they became a majority of the students? It would severely test the credibility of the principle in practice and ultimately force a reformulation of the principle.

One of the bedrock beliefs in public education (and in the professional reform stream) is equal treatment and support of all students. Equal treatment has historically been operationalized as an issue of equal expenditure of general operating school dollars behind each child plus additional dollars targeted to needy children. California and LAUSD have pressed fiscal equity about as far as any entity in any state. *Serrano v. Priest* (1974) significantly leveled the amount of money behind each child across school districts in the state, and *Rodriguez v. Los Angeles Unified School District* (1986) further refined this same measure across individual schools in the LAUSD. Equality of fiscal resources has served as a template to guide the flow of educational services across different school districts so that the accident of residential location would not determine the quantity (and presumably the quality) of educational resources that a student received.

This bedrock equity principle has held during a period when the sociocultural and economic backgrounds of children and families were relatively homogeneous, culturally and economically. The great majority of children came largely from families that were middle class and Anglo, and the children who came from non-Anglo, low-income families were literally in the minority. For those children in the racial, cultural, and economic minority, special programs with special funding (e.g., Title I) were created to bridge the gap.

The bedrock of equal per pupil spending may be eroding—not because expenditures have become unequal again but for two other reasons. First, the mechanism Californians put in place to ensure equal per pupil expenditures across districts (and control increases in property taxes) simultaneously disenfranchised local communities from voting on annual school budgets. With no process for voting on local budgets, communities have no annual means for registering their approval or disapproval of local school policies and practices. (Professional school reformers, eager to exhort teachers to involve parents and community members in local decisions, are silent on the issue of local voting on budgets.) What remains for school communities to exercise their voice is the local school board election, a vastly more blunt instrument. Second, because the current definition of equity has not yielded the anticipated results educationally, a new, more contemporary definition of equity may emerge—a definition that acknowledges new factors of equity that are more powerful than the old. Specifically, earlier definitions of equity were founded on an implicit assumption that the "average normal kid" was roughly the same throughout the state and region. (For the relatively few kids who were not "average normal," categorical grants and programs would be set up.) During recent decades, however, the proportion of children in poverty conditions has grown significantly, and the very concept of equal funding across the average normal kids is being challenged. Poverty is poverty—the problem is economic. It is less an issue of differences in school district wealth as it is an issue of differences in family wealth—differences that are growing rapidly. The demographic forces in Southern California education are actually multifold: enormous numbers of potential students to be served, unceasing growth in those numbers, and increasing proportions of children and youth living below the poverty line. Consider the scale of these demographic dimensions, for example, just within Los Angeles County.

With more than 9 million residents, or 3% of the entire U.S. population, Los Angeles County is the largest county in the country. About 1.5 million children, or 1 of every 30 children in the United States, attend one of the 1,700 Los Angeles County public schools, which are served by more than 58,000 school teachers working for 82 public school districts, not including about 200,000 children and their teachers in the county's private schools. (Slightly more than 40% of these children and teachers constitute the LAUSD.) This county is already home to about 3.5% of U.S. school-age children and nearly a third of California's children; enrollments are projected to continue to grow during the 1990s, bringing the estimated number of children in the county to about 1.85 million by the turn of the century. Future growth in student populations will be driven largely from live births of current residents, but many of the current residents in Southern California are first-generation residents in the United States.

Political and economic events around the globe continue to feed new immigration streams into the cultural mix of Southern California, with dozens of language groups coexisting with English as the common language that makes communication possible. Nearly one third of the residents in Los Angeles County, for example, were born outside the United States. More than half of those who were foreign-born residents came to the United States after 1980. By the year 2001, more than 1.4 million of the students in Los Angeles County will be Hispanic or Latino, less than 400,000 will be white, about 250,000 will be Asian (including Filipino and Pacific Islanders), and less than 200,000 will be black. In 1993, more than half (at least 52%) of the public school students in Los Angeles County spoke a language other than English at home. The enormous potential value of such language diversity for this region cannot be underestimated, but the potential is far from being realized. Limited proficiency in English unfortunately correlates highly with lack of success in school (Parker, 1994, pp. 1.1-3.21).

The home and family conditions that support high educational achievement are less widespread today than in the past, and their decline is manifested in a variety of indicators. The gap between rich and poor in the county is growing, "sharpening the image of a shrinking middle class and a growing, permanent underclass" (United Way of Greater Los Angeles, 1994).

Children and youth are finding themselves disproportionately on the low end. Despite enduring historic pockets of wealth and high personal incomes, for example, around Beverly Hills, Palos Verdes, and San Marino, growing proportions of children are raised in poverty conditions. Indeed, the poverty rate in the county has been growing steadily— 11% of the county population was poor in 1970 (14% for children), 13% in 1980 (19% for children), 15% in 1990, and 17% in 1992. In 1993 (in real numbers), 1,575,000 individuals were living below the poverty threshold ($14,350 for a family of four). More telling, perhaps, is that in 1990, 1 in 5 children was poor, compared with 1 in 8 adults and 1 in 11 elderly (United Way of Greater Los Angeles, 1994, p. 11).

Other poverty-related measures that focus more directly on students are rising even more rapidly than measures of the general population. These measures include numbers associated with individuals receiving Aid to Families With Dependent Children (AFDC) and participation in the federal Free and Reduced Lunch Program. Currently, almost 350,000 students in Los Angeles County public schools receive AFDC. The AFDC rate has grown each of the four years from 18.5% in 1990 to 23.5% in 1993. More than 800,000 students in Los Angeles County public schools participate in the subsidized lunch program. The proportion of children receiving free lunch has increased in each of the 4 years from 48.5% in 1990 to 56.0% in 1993. Children and teens in the county who were homeless in 1992 numbered 54,000, a homelessness rate of 2% for youth, compared with less than 1% for adults. One in every four children under 18 (592,000) received public assistance in 1993, compared with one in every seven adults (United Way of Greater Los Angeles, 1994, p. 23).

Educators, even with access to steady and increasing resources, would find it increasingly challenging to counteract the deleterious effects of changing home conditions. Even if such changes in home life were, magically, not taking place, the public and increasing proportions of educators have grown increasingly dissatisfied with the performance levels of graduates of educational institutions, especially of public schools. This nationwide dissatisfaction with school operation, crescendoing during the last 15 years, has led to a reformulation of what constitutes effective schooling practice, that is, professional reform. The inability (radical reformers might suggest unwillingness) of the education profession to educate these young people to high levels of performance has fueled the arguments of radical reformers for more dramatic changes.

Exclusively Public Expenditure for Public Education
Versus Continuous Forced Reduction in Public Fiscal Support

Professional reform is difficult to sustain in a period of relatively shrinking public resources coupled with growing inequality of household incomes between rich and poor. When publicly raised (tax) revenues become sufficiently inadequate for those who want increased schooling services and are willing to pay for them, the search for other revenues will escalate. Despite growing proportions of the poor in Southern California and public school spending per pupil that has fallen to 41st among states in the nation, the median household income of Californians is the *8th highest* in the United States, up from 10th highest a decade ago (Roberts, 1993, p. 272.) One of the core platforms of professional reformers entails relying solely on publicly raised public expenditures to support public elementary-secondary education. California has carried the principle further, defining state government as *the only* public agent that can routinely (annually) raise and spend public revenues for public schools. Local communities currently cannot vote annually on local tax expenditures for the operation of local schools. (School districts can, however, hold onetime special elections to issue construction bonds or seek voter approval to levy special purpose parcel taxes or "sin taxes" to finance school operations or special purchases.)

Professional reformers have also maintained a high mental wall between private and public schools: Private (largely parental) dollars go into private K-12 schooling, and public tax-raised dollars fund public K-12 schooling. Voucher plans that commingle public vouchers with private tuition dollars are heavily opposed by leaders of education organizations. (Despite their high visibility, private resources to public schools such as foundation grants represent an infinitesimally small fraction of public school expenditures.) What is the likelihood that publicly raised dollars alone will continue to support public education at a sufficiently high level to keep private money out? To state this differently, will the time come when revenue from tax dollars for K-12 education is perceived to be so inadequate that public plans will be created that redefine "public financial support" to include private (parental, business, and church) voluntary financial support into K-12 education? If this ever comes to pass anywhere, it is likely to occur in California, which has experienced one of the greatest relative declines in spending for schooling of any state in the country.

During the past 35 years, California schools have undergone a transformation that fundamentally changed their position relative to schools in other states in the country. During the previous "golden years," Californians were spending significantly more than their counterparts in other states on the schooling of each of their children. The amounts were not trivial: over 20% more than the national average as recently as 1960 (Lav, Lazere, & St. George, 1994, p. 20). Despite all arguments about efficiency and productivity that grew up here and elsewhere (i.e., "Throwing more money at schools won't solve the problem"), those additional dollars were providing relatively smaller class sizes, elementary school libraries and librarians, well-stocked secondary school libraries, school counselors, schools with lawns that were mowed and walls that were repaired and painted, sufficient quantities of recently published textbooks (with the guarantee that each student had a personal set), physical education and after-school sports programs that engaged children in teamwork, and even preventative and short-term primary health care.

In about a third of a century (a short time as these things go), California's support for children in public schools went into free fall, from one of the highest-spending states in the country, down through "average" during the 1980s, to a position today that places the state at about 85% of the average. In 1965, California ranked 5th in spending per K-12 public school pupil (Lav et al., 1994, p. 20; Parker, 1994, p. 1.1). The once golden state now ranks 41st in overall "unadjusted" spending per K-12 public school pupil and 38th on an "adjusted basis,"[8] ahead of only seven Deep South and another six rural states in dollars per pupil (Parker, 1994, p. 1.1).

Of these bottom 12 states in dollars spent per pupil, 11 are also ranked in the bottom third of all states in personal income earned per pupil. As Parker (1994) points out,

> Only one wealthy state in the nation can afford to pay much more for its students but doesn't: California, ranked in the top third of all states in personal income earned per pupil, is ranked in the bottom third of all states in spending per pupil. Today California is next to worst in the nation in the condition of its school buildings. (p. 1.1)

California is last among states in the percentage of K-12 school staff who are librarians and 42nd among states in the percentage of staff who are guidance counselors. It even ranks 35th among states in percentage who are school administrators and 45th in the percentage who are teachers.

Although California, relative to other states, seems to be concentrating its resources on instruction (62.9% within the state vs. 62.6% nationally), it has fewer actual teachers than most states: California is next to last in class size (more students per teacher than all states except Utah). This seeming anomaly—few teachers yet more resources devoted to instruction—is achieved because California has created a two-tiered instructional system. Although ranking 45th among states in the percentage of staff who are teachers, it also ranked 3rd among states in the percentage of staff who are instructional aides (many of them bilingual aides).

Despite a large number of highly experienced, dedicated teachers in Southern California, severe teacher shortages plague its classrooms. In Los Angeles County alone, about 4,800 teachers (8.4%) are employed on an emergency basis with minimum qualifications. Shortages in bilingual, science, math, and special education teachers are particularly acute.

These problems have a fiscal root, although they are manifested in reduced quality of services. The decline from 120% to 85% of average national support for schooling may be too abstract a concept to relate to everyday life. By bringing this example "closer to home" and comparing the support for a classroom of kids in Southern California with the support for a classroom of kids in the New York-New Jersey metropolitan area, one would see that the latter received about $75,000 in additional educational support each year. Imagine what could be done with this money in each classroom. Professional reformers understand that this level of resource scarcity impairs schooling services. It is difficult for the general public, however, to distinguish between impaired services due to lack of resources and impaired services due to lack of will or competence. Even when local communities wish to improve these services, they are powerless to raise public resources locally and have to rely on private collective mechanisms such as public school foundations or private individual initiative, including private schools or private supplemental education.

The budget problem goes beyond schools. The broad, sweeping fiscal and economic forces at work in California affect higher education in ways similar to K-12 education, although the common roots of the problems are not always apparent. Systems of public schooling and higher education are often considered separately—and for seemingly good reasons. The two sectors serve different age levels of students; one level is intended to serve all students, and the other is intended to serve about half that number; their purposes and programs are different; the forms of their fiscal support and expenditure differ in degree and sometimes in type; and the issues of governance seem to be different. Despite these apparent differences, the opposite is closer to the truth. State government plays a determinative role in both educational systems. The vast majority of students from California's K-12 system are those that attend California's higher education system, and public resources to support both come from the same taxpayers. The fiscal dimension of this interdependency between K-12 and higher education systems is especially pronounced.

Public fiscal support for physical facilities construction and for maintenance at all public education institutions in California has also been greatly reduced. Unlike the operating budgets that are balanced annually, capital budgets are rolling up increasing debts of deferred maintenance that accumulate year to year. As with the K-12 system, as tax support for universally low-priced higher education declines, greater emphasis will be placed on private sources of support and public support for scholarships for low-income students.

Systemwide Changes in Schooling Versus School Improvement Through "Projects"

Coupled with demographic and fiscal forces working against professional reform is the seemingly unalterable means by which professional reformers seek to improve schools, that is, through targeted projects. "Projects" in this context imply that schools and districts can elect to participate, usually through a competitive grant process. "Reform" of these schools and districts is essentially a mutually agreed-on agenda between the reformers and schools/districts. Arguably, schools and districts that win the opportunity to reform themselves are those most eager to undertake the activities outlined in the reform agenda. Less willing and less able schools/districts simply do not participate, and these are the majority of schools. Originally, professional reform efforts were acknowledged as targeted, pilot projects, but increasingly these efforts are labeled "systemic" or "comprehensive." Despite the change in labeling, these efforts still act like projects in that they are not, and perhaps cannot be, truly comprehensive.

Increasingly, the question being raised is this: Is there a reasonable likelihood that the current (and undoubtedly future) series of "systemic projects" (and all other initiatives supported by public education professionals) will bring about the idealized changes in public education that have been extensively delineated in countless commission reports? If and when the collective response is "not likely," the conversation about improving the education of young people will likely move from its current position (professional reform) to a more radical position.

How great an impact has professional reform had on schools in metropolitan Los Angeles and in Southern California? Precise measures don't exist, but several crude indexes do. In a survey of school reforms in Los Angeles County that was compiled as part of the county's competition for an Annenberg Challenge grant, about a half dozen professional reform projects were identified, all of which affected less than 200 of the nearly 1,700 schools in Los Angeles County (Quinn, 1994). In a similar budget request to the Annenberg Foundation, Northern California educators indicated that "at least 300 [of 1,203] schools [in the six-county region] have an active membership in one or more reform networks" (San Francisco Bay Area School Reform Collaborative, 1995, p. 3). The professional reform initiatives listed in both instances had been in existence between 2 and 10 years. If these collectively touch too few schools too superficially, the professional reform agenda will not be sufficiently implemented to make a positive impact on the public.

What about the majority of schools that are not among those falling under the rubric of this reform movement? Professional reformers believe that if sufficient numbers of schools are sufficiently improved, the pressure on remaining schools will be sufficient to make them improve as well. This presumption rests on three untested premises: (a) that all of the "sufficiencies" will come to pass, (b) that improved schools will stay improved, and (c) that the additional resources expended on the more willing schools and districts will be available and sufficient to induce less willing and able schools and districts. The additional resources in some instances *have* been significant, for example, several million additional dollars spent during 3 years to improve two schools through L.A.'s New American Schools Development project.

Another impediment to "scaling up" from eagerly participating schools to less enthusiastic schools derives from benefits conveyed to early participating schools that cannot be conveyed to other schools that join the reform at a later time. Schools in the LAUSD that volunteered to become LEARN schools were granted immunity from having to accept "must-place" teachers, that is, teachers who for various reasons are entitled to jobs in the school district but currently had no specific assignments. As more and more schools came into LEARN, the school district had fewer non-LEARN schools to which to assign must-place teachers. This particular benefit cannot be extended to all LAUSD schools, even if they all decided to become LEARN schools.

Yet even the problems associated with scaling up professional school reforms are not as fundamental as the way professional school reforms address the disconnection between schools and communities. Virtually all professional school reforms stress the importance of parent involvement but, with few exceptions, do not connect broader communities to schools in ways that have positive consequences (read incentives) for schools or for communities.

Conclusion: The Likelihood of Sustaining the Professional Reform Agenda in Southern California and the Metropolitan Los Angeles School Districts

Despite the notable initiatives among leaders of professional educators to improve schooling (professional reform) in Southern California, continued progress on the professional reform agenda is increasingly threatened by the combined growth in numbers of children and community heterogeneity, declining fiscal support for their school expenses contributed to in part by the absence of a local revenue option for a community's schools, and the marginal impact (so far) of professional school reform. Presuming that these three forces in combination erode the hegemony of professional reform over radical reform, what (if any) are the major levers that could possibly be employed in the short run to improve the likelihood of sustaining the predominance of the professional education reform agenda in California?

Certainly, the demise of radical reformers and radical reform initiatives would leave professional reform standing alone and on center stage. For this to happen, public perception of the public school system would have to improve significantly and/or the circumstances discussed above would have to reverse themselves, which is unlikely. The number and proportion of poor children are likely to continue to grow, if only because the primary source of that growth is in-state births, not immigration. The difficulties associated with poverty are inherently difficult to ameliorate, especially in the short run. Local communities will continue to differentiate themselves from each other, further eroding the viability of a "one size fits all" approach to public education.

It is somewhat more likely, but nonetheless difficult to imagine, that professional school improvement strategies will, by themselves, be other than selective and project based, inducing first the most willing and capable districts and schools to participate. On the other hand, given a sufficient increase in the number, scale, and scope of these projects and enough time for them to have a discernable positive impact on the public's perceptions of public school effectiveness, professional reform may gain relative to radical reform in the eyes of the public. For this string of events to occur in California, it would also be necessary to have in place a student assessment (testing) system that the public values and that, simultaneously, is capable of identifying improvements in student learning when they occur at the level of the individual student and comparable with other students, schools, school districts, states, and nations. This is not likely to happen soon.

By process of elimination, the primary opportunities and problems facing professional school reform are those of connection between public fiscal support for public schooling and a local community voice in public schooling policy and practice. This can be achieved by granting local communities the option of raising local public revenues to support local schools. Given the relatively large discrepancy between high private wealth and low public fiscal support for public education and growing local community heterogeneity, increased demand for higher-quality schooling will ultimately be reflected either (a) publicly, in increased autonomy by local school districts to increase tax support for local schools; (b) privately, in increased parental support of private schools; or (c) in a mixed system, that is, with a voucherlike initiative that can draw both public and private support for school children. (Another alternative, less likely than the others, is for the state to dramatically increase tax revenues to support local schools, say, up to the national average, through an overhaul of its current taxation system. For a critique of the current system and suggestions for improvement, see Lav et al., 1994.) By granting localities the option of raising additional public resources for schooling, both professional and radical school reformers would take a step toward each other. Professional school reformers would be required to be much more responsive to local community sentiment about schools than has been the case so far. Radical school reformers would be required to convince more of the local citizenry of the merits of individual reform initiatives than has been the case so far.

A local revenue-raising option brings to local communities more than a fiscal voice in schools. Revenue-raising options are, almost by definition, referenda on strategies for school improvement in localities. They become the mechanism for shifting public school decision making from state to local communities, thereby redressing one of the major unintended consequences of California's taxation system. Proposition 13, passed by California voters in 1978, radically altered not only the general taxation system but also restricted the right for localities to raise local school revenues. Even among its most ardent supporters, there is general agreement that one of the unintended, negative consequences of Proposition 13 was to shift much of the locus of government decision making from the local to the state level. Nowhere is this more apparent than in the case of schools. In place of local budgets funded locally, we have an incredibly Byzantine system of school finance in which a school district's fiscal fate is determined on the basis of obscure formulas understandable—if at all—by no more than a handful of experts. With spending decisions determined locally and revenues determined at the state level, local taxpayers are understandably at a loss as to who to blame when available resources invariably fall short of desired expenditure levels. With a funding system in which marginal revenues were raised locally, there would be no question as to where the responsibility rested.

Although a specific proposal for a local revenue option would of necessity have to be constructed as a vector among competing principles, the mechanism for bringing it into existence is straightforward. It is possible, through a revision of the California Constitution, to permit local communities to increase their tax effort to increase support for their schools. More fundamentally, such a change would require a redefinition of the concept of educational equality from equal spending for similar students to one of equal spending for similar students from communities *with equal tax effort*. In other words, equality of spending is placed within the context of equality of effort: Local communities that wished to increase their effort (regardless of property wealth in their school district) could increase support for their children and the state would guarantee equal levels of funding behind each student among school districts of comparable effort. (School finance specialists call this mechanism *district power equalizing*.)

The redefinition of *equality* to include "effort" would redirect the professional reform agenda slightly toward the radical reform agenda. Just as family effort interacts with family wealth in selections of private and public schools currently, local community effort would interact with school district wealth in determining the level of financial support of local schooling. By incorporating local voice (effort) into the concept of fiscal equality, a form of safety valve would create an outlet for communities that want more schooling services and are willing to tax themselves for it. Without that safety valve, and presuming continuing trends in demographics, public finance, and school improvement strategies, these same local communities will pursue alternative (more radical) agendas for school reform. Shifting decision making on public school revenues from state to local communities is preferable to shifting from a public to private provision of schooling. In California, especially metropolitan Los Angeles, one or the other shift is likely to occur.

Notes

1. The Coalition of Essential Schools is a national network of schools pursuing principles of education espoused by Theodore Sizer, professor of education at Brown University. Accelerated Schools is a national network of schools pursuing principles of education espoused by Henry Levin, professor of education at Stanford University. The New American Schools Development Corporation is a nationwide private school reform effort composed of 11 separate "break the mold" school design projects, involving altogether about 150 schools directly and many more indirectly.

2. The Urban Systemic Initiative is an initiative of the National Science Foundation designed to improve math and science education through "systemic school reform." It provides multimillion dollar grants to major cities including Los Angeles.

3. In 1994, Walter Annenberg announced a grant of $500,000,000 to improve America's schools. Since the announcement of the grant, specific awards have been made to major American metropolitan areas, including Los Angeles County.

4. It would be somewhat facile to characterize these two streams of reform in solely political terms, that is, liberal (professional reform) versus conservative (radical reform) or in solely organization and governance terms, that is, hierarchical (professional reform) versus market-oriented (radical reform), although some correlation exists in both cases.

5. A few individuals started out in one reform stream and then gravitated to the other. Joseph Alibrandi, president of Whittaker Corporation and early supporter of LEARN (professional reform), ceased participating to campaign for passage of a statewide voucher initiative (radical reform).

6. Attributed in an unsigned editorial, "To Move the Schools Forward" (1995, p. B6).

7. Title I of the Elementary and Secondary Education Act has provided the bulk of resources, which constitute approximately 5% of the budgets of many urban school districts.

8. "Adjusted" figures represented reported data inflated by 0.385% to factor in the effect of the state's law that pays for excused school absences. See Parker, 1994, p. 1.2.

References

Charter School Act of 1992, S.B. 1448, Calif. Education Code §§ 47600-47625 (West 1993 and Suppl. 1996).

Demonstration of Restructuring in Public Education Act (1989), S.B. 1274, Calif. Education Code §§ 58900-58928 (West Suppl. 1996).

Education Commission of the States. (1995). *Rising to the challenge: A new agenda for California schools and communities.* Denver, CO: Author.

Flanigan, J. (1995, May 10). Can business save schools? *Los Angeles Times,* pp. D1, D11.

Gerstner, L. V., Jr., Semerad, R. D., Doyle, D. P., & Johnston, W. B. (1994). *Reinventing education: En-trepreneurship in America's public schools.* New York: Dutton.

Hadley, S. (1995, January 10). At Moorpark firm, kids learn while parents work. *Los Angeles Times,* pp. B4 ff.

Lav, I. J., Lazere, E. B., & St. George, J. (1994). *A tale of two futures: Restructuring California's finances to boost economic growth.* Washington, DC: Center on Budget and Policy Priorities.

Marshall, R., & Tucker, M. (1992). *Thinking for a living: Education and the wealth of nations.* New York: Basic Books.

Olsen, L., Chang, H., Salazar, D. D., Leong, C., Perez, Z. M., McClain, G., & Raffel, L. (1994). *The unfinished journey: Restructuring schools in a diverse society.* San Francisco: California Tomorrow.

Parker, J. (1994). *The condition of public education in Los Angeles County, 1993-94.* Los Angeles: Los Angeles County Office of Education.

Pasadena Unified School District. (1993). *New directions: A blueprint for action.* Pasadena, CA: Author.

Perelman, L. J. (1992). *School's out: A radical new formula for the revitalization of America's educational system.* New York: Avon Books.

Policy Analysis for California Education. (1995). *Rebuilding education in the golden state: A plan for California's schools.* Berkeley, CA: Author.

Pyle, A. (1995, May 9). 103 schools added to L.A. district reform bid. *Los Angeles Times,* p. B1.

Quinn, J. (1994). *Reform in L.A. County schools.* Unpublished document.

Reform roadblock [Unsigned editorial]. (1995, April 26). *Los Angeles Daily News,* p. 12.

Roberts, S. (1993). *Who we are: A portrait of America based on the latest U.S. census.* New York: Times Books.

Rodrigues v. Los Angeles Unified School District, No. C611358 (Los Angeles County Sup. Ct., May 5, 1992).

San Francisco Bay Area School Reform Collaborative. (1995). *Raising reform to a regional scale: An Annenberg Challenge proposal.* Unpublished document.

Serrano v. Priest, 5 Cal.3d 584, 487 P.2D 1241 (Super Ct. for Los Angeles County, CA, 1971).

State of California. (n.d.). *Local Revenue Options for Schools.* Unpublished memorandum.

To move the schools forward [Unsigned editorial]. (1995, March 22). *Sacramento Bee,* p. B6.

United Way of Greater Los Angeles. (1994). *Los Angeles 1994: State of the county report.* Los Angeles: Author.

The Health Care Crisis

WILLIAM LOOS
Medical Director, Los Angeles County Department of Health Services

My personal concern is that when all the shuffling of resources is done, when the last ounce of inefficiency is cleansed from our system, and when the best possible public-private partnership is sealed, we will still be left with patients who are underserved by a local health system that is underfunded and by a national system that is overstretched because of our continued high-tech approach to medicine.

The current perspective on health care in 10 years is that large numbers of patients will remain without access to care and that the resources left to care for those patients will be frightfully meager. The county response is to reinvent government by increasing its collaboration with the community and increasing its efficiency. But without acknowledgment at national and state levels that access to expensive technology must be controlled, health care will continue to bust budgets. We may well see the familiar scene of patients having to pay co-pays they cannot afford, government paying rates that doctors won't accept, and patients seeking care from the safety net providers who may no longer be there.

The Health Care Conundrum

ROBERT E. TRANQUADA

As Los Angeles and the rest of the nation have watched the political dance around health care reform for the past several years with fascination and various degrees of optimism or dismay, it is relevant to ask how any such reforms would affect Los Angeles. Because the current status of health care access and delivery varies widely from one place to another in the United States, the potential effects of any given set of health care reform measures on each locale will vary considerably, as well. To assess potential effects of various possible reforms on Los Angeles, we must first examine the existing status of health care in this county of more than 9 million people. Thus, the questions are what distinguishes Los Angeles from the rest of the nation, and how might those specific characteristics be affected by various proposals for federal legislation on health care.

There are a number of significant differences between the current health care situation in Los Angeles and most of the rest of the United States. In brief, these include the following:

1. In Los Angeles, the proportion of the population that is medically uninsured is nearly double the proportion of uninsured in the United States and is nearly one and a half times that of the average for the state of California.
2. Compared with the rest of the United States, Los Angeles has a greatly advanced level of penetration of managed care programs in both private and publicly funded health care arenas.
3. Largely as a result of the emphasis on managed care, there is a substantial surplus of hospital beds and a growing surplus of specialist physicians.
4. A seriously underfunded county-operated health system, dedicated to the care of uninsured medically indigent persons, has legitimate needs to replace antiquated facilities at a time when public support for such expenditures is low.

213

5. There is a high proportion of non-European, legal immigrant population with a need for culturally and linguistically acceptable medical services.

6. An inadequately counted population of undocumented residents numbers perhaps 700,000, or nearly 8% of the county's population.

7. A 6-year-old economic recession has resisted correction up to this time, which has resulted in a poverty population that has grown disproportionately to the growth in total county population in the past decade.

To better place these issues in context with the possible impact of national proposals, the following sections will examine them more closely.

The Health Care Access Problem in Los Angeles County

The report of the Task Force for Health Care Access in Los Angeles County (Task Force, 1992), *Closing the Gap,* provides a rather stark profile of the situation in the county. Drawing on the work of Brown, Valdez, Morgenstern, Bradley, and Hafner (1991), Cousineau et al. (1994) recently reported updated figures: In 1992, 2.6 million people, of a total population of 9.2 million, in Los Angeles County were without any health insurance on any day of the year. This represents 31% of the population under the age of 65. These figures compare with a national estimate for uninsured persons of 17% and a state of California rate of uninsured persons of 23%. If one adds to the 2.6 million uninsured the 1.8 million residents of the county who are eligible for Medicaid, it is apparent that nearly 50% of the total population of the county is either uninsured or on Medicaid. This represents more than 50% of the population under the age of 65.

Merely identifying the number of medically uninsured persons carries with it the assumption that there is some detrimental effect to being uninsured. Does the lack of health insurance affect the health and longevity of U.S. populations? The answer is clearly yes, to both questions (Franks, Clancy, & Gold, 1993). The absence of health insurance is associated with higher rates of appendiceal perforation and peritonitis from acute appendicitis (Braveman, Schaaf, Egerter, Bennett, & Schechter, 1994), with increased mortality from breast cancer (Ayanian, Kohler, Abe, & Epstein, 1993) and trauma (Haas & Goldman, 1994), with unneeded hospitalization (Billings et al., 1993), and with increased morbidity from a number of conditions (Stoddard, St. Peter, & Newacheck, 1994). A lack of health insurance has a well-documented association with increased morbidity and mortality. Thus, Los Angeles County entered the current era of health care reform debate with the most severe concentration of uninsured and therefore unmet health care needs of any metropolitan area in the United States (Brown et al., 1991).

The Task Force (1992) estimated the need for doctor's office visits for the uninsured population on the basis of actual experience with similar populations who were covered with health insurance. They concluded that these 2.6 million people had a need for about 15 million doctor visits per year. The Task Force then estimated the available supply of doctor visits for the uninsured population. These included nearly 3 million visits provided

Table 12.1 The Supply of Doctor Visits for Uninsured People in Los Angeles County (in Millions per Year)

The Need		15 million
The Supply:		
County services	4.0	
Free clinics	0.35	
Private hospitals	1.2	
Private doctors	1.5	
Out-of-pocket	4.0	
Total	11.0	
Unmet Need		4 million

SOURCE: Task Force for Health Care Access in Los Angeles County (1992).

by the county Department of Health Services (DHS), about 350,000 by the 71 free and community clinics, more than a million uncompensated visits provided by the 130 private hospitals, 2.7 million uncompensated visits provided by the county's 24,000 licensed physicians, as well as nearly 4 million visits paid for out of pocket by persons who were uninsured. The most generous estimate was that these sources provide about 11 million visits per year, leaving an unmet need of at least 4 million visits per year (see Table 12.1).

A more detailed view of the uninsured population reveals that 85% are employed persons and their dependents. The age group from 18 to 29 has the highest rate of persons uninsured (42%), followed by the age cohort 30 to 44 (30%), 0 to 17 years (28%), and 45 to 64 years (22%). Lack of health insurance is inversely proportional to family income, with 46% of those with family incomes below 200% of poverty uninsured, whereas only 12% of those with incomes at or above 300% of the poverty level are uninsured. The Latino population has the highest rate of persons uninsured (46%); Asian Pacific Islanders follow at 25%, African Americans at 23%, and Anglos at 17% (Brown et al., 1991). It is not surprising to find inadequate health insurance concentrated among young people, who are less likely to believe that they will need health insurance, and among those with low incomes, who are most likely employed in personal services for minimum wages or in small businesses in which health insurance is less likely to be included as a fringe benefit. The high overall rate of persons uninsured and the high rate of employment among those uninsured in Los Angeles County are best explained by the county's employer mix. There are 220,000 employers in the county, of which 200,000 are employers of 49 or fewer employees, accounting for 28% of all the employees in the county (United Way of Los Angeles, 1994). Small employers are simply unable to afford health care insurance as a fringe benefit for their employees.

Growing Penetration of Prepaid and Managed Care Programs

In the United States, medical care has traditionally been offered under a fee-for-service system in which the doctor, hospital, or other provider is paid for each service provided.

Thus, the more services provided, the more income received. This principle was encompassed in the traditional form of indemnity medical insurance, which indemnifies individuals for fee-for-service expenses incurred in covered medical care.

Increasingly appearing—in Southern California since the early 1930s and in much of the rest of the United States since the early 1970s—is the concept of *prepaid care,* in which a monthly fee is accepted by doctors and hospitals (and other providers) in return for which all contracted and required medical services are provided without additional charge. No fees are collected for those services because they are prepaid. Under prepaid systems, providers can make profits if they can manage to reduce the number of required services (as opposed to fee-for-service providers who profit from providing more services).

Because these programs have begun to manage patient care more carefully, they are called *managed care systems.* Many such systems contract with multispecialty physician group practices. They may own hospitals, home care provider systems, long-term care facilities, laboratories, and other facilities. These systems are called *integrated systems* and are part of the managed care scene.

The principle of use control has been introduced by most traditional indemnity plans because insurers recognize that if they pay for fewer services, they can make higher profits or charge less for insurance. These indemnity plans are now also managing the care they will pay for. Thus, a whole new method (really several methods) of medical care organization has been collectively called *managed care.*

In general, managed care systems provide health care services at lower cost than standard indemnity insurance. In California, it is currently estimated that fully 90% of those insured with private insurance are insured under some variety of managed care. This compares with a level of 40% or less for most of the rest of the nation (T. Tingus, California Association of Health Maintenance Organizations, Inc., personal communication, 1994).

This phenomenon has two major effects that are relevant to the subject of this chapter. Under fee-for-service reimbursement, hospital beds are revenue centers because the more of them that are filled, the more revenue the hospital realizes. Under prepaid managed care systems, hospital beds cease to be *revenue* centers and become *cost* centers for the hospital. Under these systems, whenever a patient occupies a bed, it costs the hospital out of its already prepaid revenue to provide that service. This leads doctors and hospitals in prepaid systems, as well as those in other managed care systems seeking to lower their costs, to seek less expensive alternatives to the use of hospital beds (e.g., outpatient surgery for many conditions) and to hospitalize patients for much shorter average hospital stays.

This behavior is further encouraged by the Medicare Prospective Payment System (PPS). Since 1983, PPS has paid a fixed amount for the hospitalization of each specific condition (the so-called diagnostically related groups or DRGs), thus encouraging early discharge and shorter hospital stays. Consequently, a normal delivery of a pregnant woman that 10 years ago was a 3-day hospitalization is now an 18-to 24-hour hospitalization. Removal of the gallbladder that 10 years ago resulted in a 7- to 10-day stay in the hospital is now accomplished commonly with a 1- or 2-day hospital stay, using new surgical technology. These trends have led to greatly diminished use of hospital beds and have resulted directly in the surplus of hospital beds, which is discussed in the next section.

In addition, as health care insurers and providers become increasingly sensitized to the marketplace price competition for decreased costs, managed care plans no longer are willing to pay doctors and hospitals their usual fees and charges. Instead, new, lower discounted charges are negotiated. Hospitals have traditionally charged insurance companies and those who pay cash more than the actual costs of hospitalization. The surplus funds received have then been turned around and used to pay the real costs of free and reduced-charge care as part of the typical hospital's charitable responsibility. This is referred to as *cost shifting*. Some have referred to this as the "Robin Hood system" in which hospitals rob the rich (those with insurance and ability to pay) and give to the poor (those without insurance or ability to pay).

The University Health System Consortium (1995), increasingly alarmed about the future of academic health centers located in areas of high penetration of managed care, has studied the 80 or so U.S. market areas in which academic health centers are located. Each of the areas has been evaluated and classified with respect to the penetration of managed care.

First-generation markets are identified as those with less than 20% managed care penetration and a fragmented payer base with many small insurers, of which few control large volumes of enrollees. Second-generation markets have moderate (20% to 40%) managed care penetration and a moderately fragmented payer base. Third-generation markets have high managed care penetration (more than 40%) and a consolidated payer base in which a few large insurers control the bulk of the market. In the fourth-generation markets, there is a high (greater than 60%) penetration of managed care and a highly consolidated payer base. In this generation of markets, the majority of managed care insurance is managed on a capitation basis at a certain price per enrollee per month.

This analysis of market conditions identifies 18 market areas in first-generation status; 27 in second-generation status; and 10 in third-generation status, including Boston, Denver, Orange County, Portland, Sacramento, San Diego, San Francisco, Oakland, Seattle, and Worcester. Only two areas are considered to be in the fourth generation—Los Angeles and Minneapolis/St. Paul. It is noteworthy that of the 12 areas estimated to be at the third- or fourth-generation level of managed care development, half of them are located in California (University Health System Consortium, 1995).

Now that 90% of payers are in some form of managed care in Los Angeles, the lower discounted prices received by hospitals have greatly diminished the ability of community hospitals to cost-shift to subsidize care for uninsured patients. This reality is particularly important in hospital emergency rooms used by uninsured persons. Emergency rooms are required by state and federal law to provide medical evaluation of any patient, regardless of ability to pay. For this reason, emergency rooms are used excessively by persons who are uninsured as a major source of primary medical care.

In many private hospital emergency rooms, the proportion of "unsponsored" or uninsured patients has averaged as high as 40% of all users. In this way, the advent of managed care has put a terrible squeeze on the ability of the community hospital to afford to provide uncompensated care. One evidence of this is that in 1982, there were 103 hospital-based emergency rooms and 21 approved trauma centers within Los Angeles County. By 1992, 10 years later, the number of emergency rooms in operation was 85 and the number of trauma centers was 13 (Cousineau et al., 1994). All those closures resulted from the combination of increasing numbers of unsponsored patients using emergency rooms and decreasing rates of payment by managed care systems, which resulted in far less money available for cost shifting and which made operation of many of the emergency rooms and trauma centers no longer economically viable.

In the past several years, the reality of diminished cost-shifting ability for many inner-city hospitals has been recognized with the state and federal program of designated "disproportionate share hospitals" (DSHs). Those hospitals whose proportion of Medi-Cal patients and uncompensated care exceeds an established threshold may be designated as DSHs. The DSHs then become eligible to receive additional income on the basis of the proportion of Medi-Cal patients they care for. This income recognizes the disproportionate share of low-paying Medi-Cal and nonpaying patients these hospitals serve by virtue of their locations in underserved areas of the county.

All the six county-operated hospitals and a small number (19 of 130) of private hospitals located primarily in inner-city areas are designated as DSHs. As increases in managed care contracts and diminished income from insurance have limited their critical cost-shifting ability to pay for free care, the DSHs have become increasingly dependent on this source of revenue to remain economically viable. As we shall see, this is a key consideration in the health care system in Los Angeles County.

Surplus Medical Care Resources

Since 1983, which signaled the end of Medicare cost reimbursement for hospitals with the introduction of the prospective payment system and the use of diagnostically related groups, there has been a growing surplus of acute hospital beds in Los Angeles (see Table 12.2). Occupancy rates have steadily decreased from average rates as high as 74% in recent years to a current occupancy level of 47%. The number of private acute general hospitals in Los Angeles County has declined from 126 hospitals with 29,200 licensed beds in 1989 to 109 hospitals with 26,600 licensed beds in 1993 (Center of Health Resources, 1994).

A major contributor to this change has been the relatively rapid increase in the proportion of insured persons who are in managed care plans. This has resulted in a decrease in average length of hospital stay in Los Angeles County from 8.3 days in 1983 to 5.1 today. The hospital use rate, measured in inpatient days per 1,000 population, has changed dramatically as the penetration of managed care insurance has increased. As can be seen in Table 12.3, managed care programs have decreased hospital use by commercial and government programs by as much as 67%, and there is evidence that even further efficiencies in the use of hospital beds may be realized. These phenomena have resulted in the creation of a major surplus of hospital beds in Los Angeles.

Table 12.2 Percentage Bed Occupancy on the Basis of Licensed Beds: Private General Acute Hospitals

Calendar Year	Los Angeles County	Six Southern California Counties*
1989	51.2	51.4
1990	50.4	50.5
1991	49.9	49.7
1992	49.4	48.7
1993	47.5	47.4

SOURCE: Center of Health Resources (1994).
*Los Angeles, Orange, Riverside, San Bernardino, Ventura, and Santa Barbara Counties.

Table 12.3 Hospital Inpatient Days Used per 1,000 Population: Fee-for-Service Versus Managed Care, 1995

Insured Population	Fee-for-Service Insurance	Managed Care Insurance
Under age 65, commercial	640	213
Under age 65, Medicaid	901	407
Medicare	3,193	1,013

SOURCE: Lewin-VHI, Inc. (1995, pp. 15-16).

At the same time, as managed care plans have gone far to remove the physician incentives inherent in fee-for-service practice and have substituted capitation and the assumption of risk by physicians, the demand for specialized physicians has begun to decrease. Yet the medical schools and residency programs in the United States and in Southern California continue to produce specialty physicians at an undiminished rate. Los Angeles is beginning to have a growing surplus of specialty-trained physicians.

Efficient medical care systems such as Kaiser Permanente provide physician staffing levels of about 60 primary care doctors (those in family practice, general internal medicine, and general pediatrics) and about 60 of all other specialists per 100,000 insured persons (Kaiser Permanente Southern California Regional Headquarters, personal communication, July 1994). This means that an efficient physician staffing level is about 120 physicians per 100,000 insured persons. The present supply of licensed physicians in Los Angeles County is about 24,000 for a population of 9,100,000—a ratio of about 264 physicians per 100,000 population, of which about 60 per 100,000 are in primary care. Because Kaiser Permanente cares for a population with a small proportion of older persons and one that is somewhat healthier than the average for the whole county, it is not practical to suggest that those exact ratios could be used.

Nevertheless, for the medical care industry as a whole to achieve even a 50% efficiency, compared with fully capitated, managed care programs, it would require about 160 physicians per 100,000 (see Table 12.4). This indicates that there would be a surplus of

Table 12.4 Physician Supply in Los Angeles County: Existing and Required Under Predominant
              Managed Care Penetration

Type of Physician	Current Supply, Physicians/100,000	Kaiser Permanente Staffing Level, Physicians/100,000	Requirement of 90% Penetration of Managed Care
Primary care	60	60	70
Other specialist	200	60	90

SOURCE: State of California Medical Board and Kaiser Permanente Southern California Regional Headquarters (personal communications, 1994).

more than 9,000 physicians in the county, almost all of whom would be practitioners of specialty medicine.

The result of these determinants of a hospital bed surplus is an increasing rate of formation of multihospital systems, hospital mergers, and hospital closures. At the same time, multispecialty physician groups are rapidly growing, with increasing buyouts of individual and group medical practices by hospital systems and physician groups as physicians seek more efficient practice modes and greater control of their future. The formation of new, independent, solo, private physician practices has, for all practical purposes, come to an end in Los Angeles. Virtually all newly trained physicians are entering some form of group practice, most commonly on a salaried basis. Compensation for primary care physicians, who are in relatively short supply, is growing relative to compensation for specialists, reflecting the increased demand for primary care physicians in managed care plans (Medical Group Management Association, 1993).

Underfunded Public Health Care

Since statehood in 1850, California has depended on the county as the political unit responsible for relief of indigent persons, including health care. This tribute to the British Elizabethan poor law has been codified since early in this century. In 1937, Section 2500 of California's Welfare and Institutions Code formalized the obligation of the counties to provide access to health care for medically indigent persons. Now, under Section 17,000 of the 1965 Welfare and Institutions Code, each county is charged by law to "relieve and support all . . . poor, indigent persons, and those incapacitated by age, disease or accident," when these persons are not "supported and relieved by their own means, or by state hospitals or private institutions" (see Tranquada & Glassman, 1992).

In response to these expectations from the state, Los Angeles County has operated its own hospitals since the 1877 founding of the Los Angeles County Hospital and Poor Farm, which was built at the site of the present Los Angeles County-University of Southern California (USC) Medical Center. By 1972, when the departments of hospitals, public health, and mental health were merged into a single Department of Health Services, the county operated more than 6,000 hospital beds in nine hospitals.

The adoption of state Proposition 13 in 1978, along with the decreasing ability of the state to meet its health care budgetary obligations, has led to a steady decline in the number of county hospitals and staffed beds. The Los Angeles County-USC Medical Center, which operated 2,100 beds in 1970, today staffs only 1,200 beds. The total number of staffed county beds has dropped to about 3,000 in six hospitals, all of moderate to advanced age. The youngest physical structure, Olive View Medical Center, was completed in 1983; the major portion of the oldest, the Los Angeles County-USC Medical Center, is a 1924 design first occupied in 1932.

In addition, the DHS operates some six comprehensive health centers that offer primary care and most types of specialty ambulatory care and 40 public health clinics scattered about the county, which provide mostly public health services and modest amounts of primary care. The clinics have been hard hit by a succession of operating budget reductions since the late 1970s and, because they are least able to generate revenues, have had to reduce their services significantly.

Total hospital days provided have decreased from more than 1,800,000 in 1970 to 1,063,000 in fiscal year 1993 (Center of Health Resources, 1994). The county property tax contribution to support the DHS was 18% of the health services budget in 1982. Since 1982, the county poverty population has grown at double the rate of general population growth. Yet in fiscal year 1994, the county contribution was only 12% of the total operating budget of $2.4 billion, and the county contribution to health services, measured in constant dollars, had declined since 1982.

In actual fact, the DHS has become increasingly dependent on revenues it receives from continued state and federal grants as the result of its enormous contribution of care to medically indigent persons. In fiscal year 1994, the inpatient population included 60% Medi-Cal, 6% Medicare, 10% self-pay and insurance based, and 24% medically indigent. Thus, any major reduction in Medi-Cal and disproportionate share revenue to the DHS, such as might occur with federalization of Medicaid (a part of a number of health care reform proposals) in the absence of implementation of universal entitlements, would severely affect the county's ability to finance the care of medically indigent persons (DHS, personal communication, 1995).

This is particularly a problem with respect to the care presently provided to undocumented residents. Although records are not kept by the DHS, it is likely that a significant proportion of county-provided services are presently used by undocumented persons. None of the proposals for national health care reform have included provisions for undocumented persons. As can be seen in the subsequent section, this poses a major threat to the county of Los Angeles.

Another fragile asset is the physical facilities that make up the county hospital stock. The DHS has a major proposal before the county board of supervisors that would result in the replacement of the oldest beds at Los Angeles County-USC Medical Center and the

Table 12.5 Ethnic Composition of Los Angeles County

Ethnic Group	Percentage of Los Angeles County Population		
	1980	*1990*	*Projected 2000*
Anglos	53	41	28
Latinos	27	37	49
African Americans	13	11	10
Asian/Others	7	11	14

SOURCE: Los Angeles County Department of Health Services (1988).

redistribution of needed hospital capacity to existing county hospitals in Torrance, Willowbrook, Sylmar, and Lancaster. This $2.4 billion project is planned to be financed from operating revenues, a substantial state contribution, and operating savings resulting from more efficient operation. In addition, because major damage to county hospital facilities occurred in the January 1994 earthquake, substantial Federal Emergency Management Agency (FEMA) contributions to reconstruction are also anticipated.

The current occupancy rate of 47% in the private hospitals in the county, accompanied by fierce economic pressures generated by increasing managed care (see previous sections), has encouraged many private hospitals to oppose the rebuilding of county facilities. They prefer that the county contract its patient care responsibilities to private hospitals, claiming that would be less expensive. The reality is that because the county earns major revenues by virtue of the disproportionate share status of its hospitals, the bulk of this revenue would almost certainly be lost if county-operated hospitals were to close. Services to indigent persons, for which the county provides about $850 million per year (about $300 million in county appropriations plus about $500 million in disproportionate share and related revenues) would be reduced to only $300 million (DHS, 1995-1996 budget presentation, 1995). Neither the county nor the private sector could begin to provide the requisite amount of care under those circumstances. It is a Catch-22 situation.

The key question remains whether any national health care reform will be legislated, now or later, by Congress, and if so, what effect it might have on the remaining obligation of the county to provide health care and on the revenues available for that purpose.

A Diverse Population

The 1990 census revealed continuing change in the ethnic makeup of the Los Angeles County population (see Table 12.5). The "majority" Anglo population now constitutes only 44% of the total, whereas the Latino proportion has grown to 38% and the Asian Pacific Islander proportion to 7%; the African American population has stabilized at about 11%. The significance of these figures for health care resides in the rapidly growing numbers of people whose first language is not English and whose cultural backgrounds, perceptions of health care, and disease vary significantly from that of the acculturated native population.

With this reality, the need for linguistically and culturally appropriate medical services has grown greatly and is severely challenging the resources of all providers of medical services, both public and private. More significant, those who are monolingual non-English speaking and those whose cultural background inhibits taking full advantage of existing medical services are concentrated among those with lower incomes who qualify for publicly supported services. Thus, the strain of caring for diverse populations is disproportionately felt by public services and further burdens an already inadequate system.

Undocumented Population in Los Angeles

Efforts to provide an accurate count of undocumented residents in Los Angeles County are frustrated by the very nature of the issue. Such individuals are not likely to identify themselves, for obvious reasons. A study by the Los Angeles County chief administrator's office estimated that there were as many as 700,000 such individuals in Los Angeles County (Internal Services Department, 1992). Although no completely reliable estimates exist, it seems highly probable that as much as 90% of this population have no health insurance, that their average income is at or near poverty levels, and that their presence has not been fully recorded in 1990 census figures or in estimates of the number of uninsured persons. Thus, the real population of Los Angeles County is probably in excess of 9.3 million, rather than 9.1 million; the proportion of uninsured persons, if undocumented persons are included, is closer to 34% of those under 65, rather than 30%.

The political liability of the great burden of undocumented residents in Los Angeles is important to understand while debate about national reform of health care continues. Five states—California, New York, Florida, Texas, and Illinois—are believed to harbor the vast majority of all undocumented persons in the United States. Because this is a burden of little significance to nearly all the rest of the states, those states' concerns are hardly focused on the issue. This means that efforts to provide relief at the federal level will find natural support in only 10 senators of the highly affected states and the 20% or 25% of the total representatives who come from these states, if they all agree on any measure. That is not a sufficient concentration of political power to guarantee success of federal relief measures for these states and their burden.

The situation with respect to undocumented people in Los Angeles is even further clouded by the success of state Proposition 187 in November 1994. Provisions of this proposition would eliminate all public funding for the health care of undocumented persons except emergency care and would require health care workers to report such individuals to the state and to the Immigration and Naturalization Service. To date, the courts have delayed the implementation of the provisions of Proposition 187 until the legal challenges against it can be resolved—which may well take until even beyond 1996. It does not change the reality that none of the health care reform proposals introduced through the end of 1994 included any provision for undocumented persons. But it does pose potentially stunning questions that range all the way from basic ethical questions relating to the responsibility of various providers to provide care to pragmatic questions about control of contagious disease and essential prenatal care for infants who will be U.S. citizens and therefore eligible for publicly supported medical care when they are born (Ziv & Lo, 1995).

A Prolonged Economic Recession
and a Growing Poverty Population

Since 1988, there has been a prolonged recession in Los Angeles (and Southern California), which began to show signs of relief by late 1994. This has been accompanied by a progressive loss of more than 200,000 jobs in aerospace and other manufacturing industries since 1988. The unemployment rate, which had been slightly below the national rate in 1989, has been substantially higher than the national rate since then and has not participated in the national declines of the past 2 years, until just recently. The county's poverty rate has steadily increased from 10.7% in 1970 to 17.2% in 1992. This may reflect the loss of well-paid manufacturing jobs, increases in low-wage work, and the growth of part-time and temporary work (United Way of Los Angeles, 1994).

These negative factors affect the health care situation in Los Angeles by adding to an already high incidence of medically uninsured persons as the number of unemployed, and therefore uninsured, persons, grows. The resulting labor surplus reduces pressure on small employers to purchase health insurance for their employees. In turn, with lower average wages, uninsured workers are unable to purchase their own insurance. These factors further increase the burden on the safety net institutions (free clinics and charitable and public hospitals), impairing their future viability.

The Effects of Proposed Reform
Measures on Los Angeles County

The preceding sections have characterized the issues confronting health care systems in Los Angeles County. I now turn to the most probable effects of several elements of proposed health care reform on Los Angeles.

HEALTH INSURANCE REFORM

Much has been written recently about proposed measures of insurance reform with which many congressional members seem to agree. These generally include such requirements as forbidding the cancellation of insurance when illness occurs, community rating of health insurance, the discontinuation of exclusions of coverage for preexisting conditions, and portability of health insurance from one place or status of employment to another. These measures are responsive to a general agreement that the availability of health insurance should be guaranteed and that it should not be cancelable for virtually any reason. The question is what effect such measures would have on Los Angeles, where employment is dominated by small businesses in which 200,000 of the 220,000 employers have less than 50 employees, only 40% of whom have health insurance (United Way of Los Angeles, 1994).

Much of the above reform has already been largely accomplished for small businesses in California through the Small Employer Group Health Coverage Reforms (A.B. 1672) of 1992 (California Association of Health Maintenance Organizations, Inc., 1992). The Health Insurance Plan of California, established 2 years ago by A.B. 1672, was created as a state agency responsible for bringing small employers together into larger groups and then overseeing the purchase of less expensive health insurance for the larger groups. The disappointing aspect about the availability of health insurance to small employers for reasonable cost is that few small businesses have taken advantage of the availability of this insurance, and a significant number of those who have, did so just to exchange existing and more expensive or more restrictive insurance for insurance of lesser cost. The result has been a disappointingly small number of new policies purchased by small employers statewide. An extrapolation of this behavior suggests that such a measure implemented at the federal level would do little to ease the situation of uninsured persons in Los Angeles. Even with such reform, as long as the purchase of insurance by small employers remains voluntary, employees who are now uninsured are not likely to have enough money to take advantage of the availability of insurance.

An additional problem, should insurance reform include the requirement for the provision of community rating to all health insurance, would be the effect of adverse selection in increasing the price of insurance. That is, if insurance were guaranteed to all at community rates, many healthy young people and some others more at risk for illness would wait until they were ill before purchasing insurance. That would tend to concentrate sick people among the insured population, which would increase the number of services required by the insured and therefore drive up the cost of health insurance. It is possible that such cost increases could lead to the abandonment of existing insurance by small businesses and others who may currently supply insurance to their employees. The result could be a paradoxical increase in the number of uninsured persons in Los Angeles County. Even if wildly successful, insurance reform would make only a small dent in the current number of uninsured persons in the county.

FEDERALIZATION OF MEDICAID

Several proposals would federalize Medicaid. These proposals include making a larger proportion of the medically indigent population eligible for Medicaid benefits and making the eligibility rules for Medicaid uniform across the states. Presumably, this would relieve the state and the counties of some considerable current Medicaid (called Medi-Cal in California) financial burden. If such a step were taken and there were an end to disproportionate share payments to county and private disproportionate share hospitals, or if those now insured by Medi-Cal could choose to move from the use of DSHs to others of their choice, the result could be disastrous for the safety net hospitals, both private and public, in the county. DSHs are now so dependent on DSH payments to meet their cost-shifting needs that loss of DSH status or of significant DSH funding to other sectors or other hospitals would end the remaining capability of both public and private hospitals to meet the needs of the uninsured.

SINGLE-PAYER PROPOSALS

Single-payer proposals generally would impose a payroll or other tax for health care, collected by a state or the federal government. The state or federal government then becomes the insurer for all the citizens in the jurisdiction. The most often referenced single-payer system is the Canadian system, in which every resident has a Medicare card, the province pays the charges of private physicians for all authorized health care services, and both private and public hospitals are financed by annually negotiated global budgets. In November 1994, such a proposal (Proposition 186) was defeated in the California general election.

A single-payer system with "universal" coverage, adopted nationally or at the state level, would eliminate almost entirely the requirement for maintenance of a publicly sponsored safety net, particularly because most such proposals include provisions that would confiscate all existing state and county funds presently used for safety net purposes. The effects on private hospitals would be profound. Little or no uncompensated care (except for undocumented persons) would be required, and virtually all services would be compensated. Private hospital global budgets would be negotiated annually, except for hospitals that are capitated providers, in which case they would negotiate with the integrated system for their rates.

HEALTH INSURANCE PURCHASING COOPERATIVES

For all practical purposes, these already exist in California. The California Public Employees Retirement System board already negotiates successfully with insurers for the insurance and managed care coverage of a million persons. The Health Insurance Plan of California, discussed previously, is another prime example.

Such a mechanism, with its centralized purchasing power for the vast majority of lives to be insured, would undoubtedly result in further pressure on provider income, which might be expected to decrease even further the margins available for cost shifting to support the uninsured. Thus, purchasing alliances in the absence of universal coverage could severely affect access to care of uninsured persons. It seems unlikely that this mechanism could lead to any major further increase in the percentage of those covered by some form of managed care in Los Angeles. We are already nearly fully engaged in that.

EMPLOYER MANDATE

The imposition of an employer mandate or an individual mandate would have a profound effect on access in Southern California. Because 87% of those uninsured are employed or are the dependents of the employed, an employer mandate by itself, without any subsidies for the poor, could reduce the number of uninsured persons from 2.7 million to just 350,000, or only 4% of the population under 65, and nearly all of those would be among the unemployed (Brown et al., 1991). This would also decrease the burden of cost shifting on providers to a level that would be almost tolerable.

NONMANDATORY SUBSIDIES FOR
PURCHASE OF PRIVATE HEALTH INSURANCE

None of the programs that are currently proposed, including subsidies for the voluntary purchase of health insurance by low-income people, seem to make much of an impact on the medical care economy. These plans provide that the purchase of health insurance would remain voluntary and that individuals who fall below certain income standards would be able to purchase subsidized policies. For those whose incomes are less than twice the poverty level ($24,000 for a family of four), however, it seems unlikely that any significant number would have the resources to pay the balance required. Such measures appear to offer little relief for the situation in Los Angeles. The cost of living is just too high for this to be expected to increase health insurance purchases measurably.

VOLUNTARY MEDICAL SAVINGS ACCOUNTS

Proposals for voluntary medical savings accounts generally include provision of a before-tax annual contribution to a so-called health savings account. Unused funds in the account would accumulate interest tax free, and the account could be used either to purchase health insurance or to pay for specific medical care costs as a type of self-insurance fund. The maximum annual before-tax contribution would be capped at some level. This is a simple idea based on the belief that people who are allowed to save money tax free for health care expenses would do so in large numbers.

Such a program, however, could not be expected to make any significant impact on the situation in Los Angeles. Young people with reasonable income can now receive tax deductions for the purchase costs of health insurance if it exceeds 7.5% of adjusted gross income. Those who choose not to are betting that they will not get sick. There is little reason to believe that this behavior will change even with full tax deductibility of these accounts. Persons who are poor and near poor (up to 200% poverty-level income) do not have discretionary resources with which to finance such accounts, and the middle class is largely covered with employer furnished health insurance already.

Conclusion

It seems clear that the unique combination in Los Angeles of small-employer dominance of the labor market and substantial numbers of medically uninsured persons when compared with the rest of the nation will require particular measures for federal legislation to significantly reduce the number of medically uninsured persons. The few popular and politically relatively neutral proposals that have a fairly good prospect of success in the Congress (health insurance reform, federalization of Medicaid, nonmandatory subsidies for purchase of private health insurance, and voluntary medical savings accounts) would not, either separately or together, have significant effects on the number of uninsured persons in Los Angeles County. The only remedy that could make a substantial impact on the situation is one that includes some variety of universal or nearly universal access.

The combined effects of a measure that incorporated the now defeated Clinton proposals, or some close variant of them, including essentially universal access, cost controls, health insurance reform, some degree of tort reform, and dynamics favoring the growth of managed care, would positively affect Los Angeles County. Universal access, without regard to the specifics of its method of funding, would come close to solving the major access problem and would remove the dreadful issue of diminished cost shifting to care for unsponsored patients. Cost controls, tort reform, and nearly universal application of managed care are, for all practical purposes, already a part of the scene in Los Angeles.

Continuing would be the problem of undocumented persons, which has not been included in the details of any of the national proposals. This problem remains to be solved. It is conceivable that the broad availability of at least a minimal health care policy that might provide coverage for six doctor visits, immunizations, and a maximum of six hospital days per year, at a cost of perhaps $60 or $70 per month, might be attractive enough to solve a large proportion of the problem.

The overriding fact is that Southern California, most particularly Los Angeles County, has some unique challenges that are simply not shared with the rest of the nation and are only partially shared by the rest of the state. It is critical that people understand them as we evaluate and make our input into the many proposals at all levels of government that may affect them. Clearly, only a comprehensive form of state or national health care reform that includes some type of universal coverage will provide significant benefit to the particularly complex situation in Los Angeles County.

References

Ayanian, J. Z., Kohler, B. A., Abe, T., & Epstein, A. M. (1993). The relation between health insurance coverage and clinical outcomes among women with breast cancer. *New England Journal of Medicine, 329,* 326-331.

Billings, J., Zeitel, L., Lukomnik, J., Carey, T. S., Blank, A. E., & Newman, L. (1993). Impact of socioeconomic status on hospital use in New York City. *Health Affairs, 12,* 162-173.

Braveman, P., Schaaf, V. M., Egerter, S., Bennett, T., & Schechter, W. (1994). Insurance-related differences in the risk of ruptured appendix. *New England Journal of Medicine, 331,* 444-449.

Brown, E. R., Valdez, R. B., Morgenstern, H., Bradley, T., & Hafner, C. (1991). *Health insurance coverage of Californians in 1989.* Berkeley: University of California.

California Association of Health Maintenance Organizations, Inc. (1992). *Small employer group health coverage reforms, A.B. 1672 (Margolin), Chapter 1128, Statutes of 1992.* Sacramento, CA: Author.

Center of Health Resources. (1994, December 16). *Quarterly Census Report.* Los Angeles: Author.

Cousineau, M. R., Ng, L., Pitts, D., Shen, S. M., Wyn, R., Kominski, G., & Brown, E. R. (1994). *At risk: Los Angeles County, the health of its people and its health system.* Los Angeles: UCLA Center for Health Policy Research.

Franks, P., Clancy, C. M., & Gold, M. R. (1993). Health insurance and mortality: Evidence from a national cohort. *Journal of the American Medical Association, 270,* 737-741.

Haas, J. S., & Goldman, L. (1994). Acutely injured patients with trauma in Massachusetts: Differences in care and mortality, by insurance status. *American Journal of Public Health, 84,* 1605-1608.

Internal Services Department. (1992, November 6). *Impact of undocumented persons and other immigrants on cost, revenues and services in Los Angeles County.* Los Angeles: Los Angeles County Office of the Chief Administrative Officer.

Lewin-VHI, Inc. (1995, January). *Report to the steering committee for the study of Los Angeles health resources.* Sausalito, CA: Author.

Los Angeles County Department of Health Services. (1988). *Los Angeles population estimation and projection system.* Los Angeles: Author.

Medical Group Management Association. (1993, September). *Physician compensation and production survey.* Washington, DC: Author.

Small Employer Group Health Coverage Reforms of 1992 (California A.B. 1672), Chap. 1128, Stat. of 1992.

Stoddard, J. J., St. Peter, R. F., & Newacheck, P. W. (1994). Health insurance status and ambulatory care for children. *New England Journal of Medicine, 330,* 1421-1425.

Task Force for Health Care Access in Los Angeles County. (1992, November 24). *Closing the gap: Report to the Los Angeles County Board of Supervisors.* Los Angeles: Author.

Tranquada, R. E., & Glassman, P. A. (1992). Providing health care for the uninsured and underinsured in Los Angeles County. In J. B. Steinberg, D. W. Lyon, & M. E. Vaiana (Eds.), *Urban America: Policy choices for Los Angeles and the nation* (pp. 312-318). Santa Monica, CA: RAND.

United Way of Los Angeles. (1994). *Los Angeles 1994: State of the county report.* Los Angeles: Author.

University Health System Consortium. (1995). *Market evolution model.* Oakbrook, IL: Author.

Ziv, T. A., & Lo, B. (1995). Denial of care to illegal immigrants: Proposition 187 in California. *New England Journal of Medicine, 332,* 1095-1098.

Remaking the City

MARGARET CRAWFORD
Southern California Institute of Architecture

A massive restructuring of urban space is occurring in Los Angeles right now. Restructuring is not the same as disintegration, even though it may appear that way to those of us in the physical side of planning. The physical structure of the city is changing in powerful ways. This poses a series of both positive and negative prospects and difficulties. We shouldn't look to the past for solutions. This leads to nostalgia and confuses our reading of the present. We need to restructure our thinking before we impose a new set of normative standards. We need to look at what's going on and *try to understand cities in new ways.*

I would ask, are cities subjects or objects? Is the city an actor with its own destiny that can be realized, an entity with a special mission? Or is it an object fought over and shaped by a variety and range of different groups and interests? I believe it is the latter. Given this, identifying the actors is a critical task for discussing a city's development. Who actually makes decisions? Is it an amorphous group of people? Who do they represent? How are these decisions implemented?

These are the old and familiar questions about planning, power, and politics that need to be asked before we propose any visions for Los Angeles.

Transporting Los Angeles

GENEVIEVE GIULIANO

Introduction

Los Angeles is known throughout the world as the prototype of the late 20th-century city. Its extensive mix of low- and medium-density communities distributed over more than 3,500 square miles and connected by hundreds of miles of high-capacity expressways represents the essence of a metropolitan structure developed around the private automobile. Those who are familiar with the history of Los Angeles understand that the dispersed form of the region was to a great extent the product of major infrastructure investments made at the beginning of this century, before the adoption of the automobile. The interregional rail system and the water distribution system made development possible throughout the region and set the stage for the multicentered, transport-intensive metropolis that Los Angeles has become.

The extensive highway transportation system in Los Angeles is both a great strength and a great weakness. The transportation system has allowed the region to grow at a phenomenal rate, and it has given most of its residents and businesses a level of mobility unparalleled in the world. The private automobile is also a major source of air pollution, congestion costs billions of dollars in lost productivity annually, and households without cars are denied the opportunities that "automobility" provides. This chapter examines L.A.'s transportation issues. I will begin by presenting some information on travel trends

AUTHOR'S NOTE: Portions of this chapter were adapted from "Transportation Policy Options for Southern California," a paper presented by M. Wachs and G. Giuliano at the symposium "Policy Options for Southern California," November 1992, Los Angeles, CA.

Table 13.1 U.S. Private Automobile Registrations per Population

Year	Auto Registrations[a]	Population[b]	Autos per Population
1910	0.46	92.0	1 per 200
1930	23.0	122.8	1 per 5.3
1950	40.4	150.7	1 per 3.7
1970	88.8	203.2	1 per 2.3
1990	142.4	248.7	1 per 1.7

SOURCE: Altshuler, Womack, & Pucher (1979), p. 24; Federal Highway Administration (1990), pp. 16-17.
a. In millions; includes private and commercial automobiles.
b. In millions.

and urban patterns relevant to transportation policy and then discuss the current situation in Los Angeles.

Travel Trends and Urban Patterns

Dominance of the private automobile is not unique to Los Angeles. In a continuation of trends that have been in evidence for several decades, the most recent national survey data (1990) show that throughout the United States, people own more private vehicles, use them more frequently, drive more miles, and are more likely to drive alone than ever before. Here are some examples. Table 13.1 gives private auto registrations per population for the United States. There were 460,000 autos registered in 1910, or 1 for every 200 people. The most rapid increase in ownership took place in the 1920s. The Depression and World War II slowed the trend, but auto ownership has continued to increase through 1990. If privately owned light trucks (minivans and compact trucks are in this category) are included in 1990 registrations, the national figures are 179.8 million private vehicles, 191.4 million persons of driving age, and 167 million licensed drivers. Los Angeles has historically had a higher rate of auto ownership than that of the United States; the ratio of autos per population reached 1 per 3 by 1929 and 1 per 1.7 in 1979. The ratio, however, has remained stable since then; thus, the 1990 ownership rate in the region is the same as that of the United States.

Rising car ownership is also illustrated by the decrease in the number of households with no vehicles and the increase in households with more than one vehicle. U.S. census data show that about 21% of all households had no vehicles in 1960; by 1990, the percentage dropped to 11%. In contrast, the share of households with three or more vehicles increased from 2.5% in 1960 to 17% in 1990 (Rosetti & Eversole, 1993). The 1990 figures for the L.A. region are about 7% with no vehicles and 21% with 3 or more (Vincent, Keyes, & Reed, 1994).

Observed increases in private vehicle travel during the past decade have been far in excess of population or employment growth. Between 1983 and 1990, private vehicle

Table 13.2 Los Angeles and Large Metropolitan Area Travel Characteristics, 1990

	All Metro Areas With > 2 Million Population	Los Angeles
Percentage of households with > 1 vehicle per driver	76%	82%
Share of person trips by mode, all vehicle trips		
Private vehicle driver	61.1%	63.4%
Private vehicle passenger	21.4%	23.9%
Transit	3.7%	1.7%
Walk	10.6%	8.3%
Other	3.1%	2.7%
Private vehicle trip length, by trip purpose (in miles)		
To work	11.4	12.3
To shop	4.1	3.5
Social or recreational	9.6	9.2
Persons per vehicle		
Work trips	1.12	1.20
Nonwork trips	1.63	1.64

SOURCE: Vincent, Keyes, & Reed (1994).

miles traveled (VMT) increased 37%, whereas population increased by just 4%. Growth in VMT reflects increases in the number of trips, longer trips, more trips by private vehicle, and more driving alone (Vincent et al., 1994). In contrast, public transit use has continued to lose market share and now accounts for just 2% of all person trips and 5.5% of all journey-to-work trips (Hu & Young, 1992).

Comparing travel characteristics of the 12 U.S. metropolitan areas with more than 2 million population shows that vehicle use patterns in Los Angeles are somewhat different from to those of other large metropolitan areas (Table 13.2). The level of car ownership is notably higher; more trips are made by private vehicle, and fewer are made by transit and walking. For work trips, however, Los Angeles has the highest vehicle occupancy, meaning that more people are being carried in these vehicles. And although work trips are longer, other types of trips are shorter than the average for large metropolitan areas.

Changes in metropolitan form have proceeded in concert with an auto-oriented transport system. Both population and employment have decentralized. Figures 13.1 and 13.2 illustrate these trends for the L.A. metropolitan region, by county. Figure 13.1 shows population growth by county for the decades since 1960. The entire area has experienced dramatic growth, but population increased most rapidly in the suburban counties. Figure 13.2 shows similar trends for employment for the decades since 1970. Jobs continue to be somewhat more centralized than population in Los Angeles County but nevertheless are rapidly shifting to suburban counties.

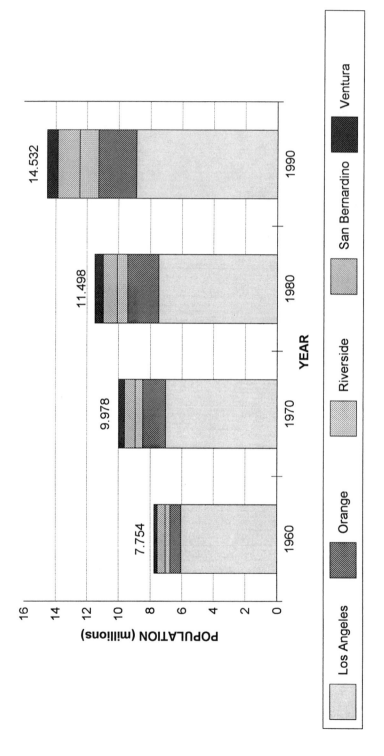

Figure 13.1. Population Shares by County, Los Angeles Metropolitan Region
SOURCE: U.S. Bureau of the Census.

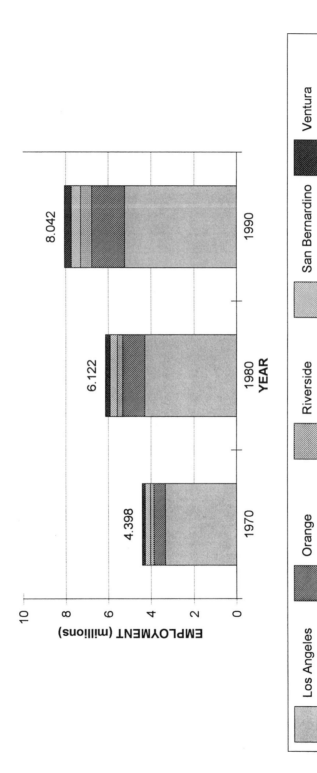

Figure 13.2. Employment Shares by County, Los Angeles Metropolitan Region

SOURCE: U.S. Department of Commerce.

Decentralization has been accompanied by dispersion; more employment growth has occurred outside major employment concentrations. To illustrate, I define as an *employment center* any area with a total of 20,000 or more jobs and with a density of at least 20 jobs per acre. In 1980, there were 10 such centers in the Los Angeles region, and these centers contained about 17% of the region's employment. In 1990, there were 12 centers, but these 12 centers contained just 14.4% of regional employment. Most travel is oriented to dispersed, low-density destinations, and there are no clear patterns of travel flows. It is important to note that decentralization of population and employment has not resulted in longer trips (Table 13.2); households are able to work, shop, and so on relatively close to home because of this intermixed land use pattern.

Transportation Problems

The region's transportation problems include traffic congestion, air pollution, and inadequate mobility for persons who are transportation disadvantaged.

TRAFFIC CONGESTION

It is difficult to estimate the extent of congestion, but the few studies that have been done show that Los Angeles is one of the most congested metropolitan areas in the United States (Hanks & Lomax, 1991). Many portions of the expressway system operate under congested conditions several hours per day, and some expressway segments have a "peak period" that lasts 12 or more hours each day.

Growing traffic congestion is the outcome of rapid population and employment growth accompanied by even more rapid growth in private vehicle travel while highway supply has remained essentially fixed. Absent major changes in travel behavior and transportation system supply, congestion is expected to greatly increase. Current regional forecasts, for example, show the time lost due to traffic congestion increasing from a 1990 base of 2.1 million hours per day to 15.5 million hours per day in 2015.[1]

AIR POLLUTION

Perhaps more serious is L.A.'s air pollution problem. The U.S. Environmental Protection Agency rates air pollution in Los Angeles as the worst in the United States. In 1989, one or more of the federal standards were exceeded on 219 days (South Coast Air Quality Management District [SCAQMD], 1991). Photochemical smog is the most important urban air pollution problem. Smog is generated from chemical reactions between hydrocarbons (HC, also known as reactive organic gases, or ROG) and oxides of nitrogen (NO_x). The products of this chemical reaction include ozone (O_3) and carbon monoxide (CO), both of which are known to negatively affect human health (Calvert, Heywood, Sawyer, & Seinfeld, 1993). Motor vehicles also emit oxides of sulfur (SO_x). Sulfur dioxide (SO_2)

Table 13.3 Emissions Shares by Major Source, South Coast Air Basin, 1987

Source	ROG	NOx	CO	SOx	PM10
Stationary	49.96%	24.09%	1.99%	37.31%	93.58%
Mobile	50.04%	75.91%	98.01%	62.69%	6.42%

SOURCE: South Coast Air Quality Management District (1991).
NOTE: Based on average daily emissions in tons per day.

contributes to particulate formation and acid rain. Finally, motor vehicles (particularly diesel-fueled vehicles) emit particulates (PM_{10}), which are irritants to the respiratory system.

Vehicle emissions account for a significant portion of each of these pollutants. The transportation sector generates about 80% of U.S. CO emissions, 45% of NO_x, and 36% of ROG (Environmental Protection Agency, 1993). The contribution of the transportation sector to air pollution differs by region. Table 13.3 gives the share of each major air pollutant that is generated by surface transportation sources for the South Coast Air Basin. The ROG share is likely underestimated; recent research suggests that actual vehicle emissions are about twice as great as previously estimated (Small & Kazimi, 1994).

The traditional response to pollution has been regulatory and technology based. Pollution controls were implemented in Los Angeles in the 1960s. Automobile emission control systems were mandated in 1973, and dramatic reductions in emissions rates have been achieved. Air quality in Los Angeles has improved as a result of these policies; 1991 was the "cleanest" year for all major pollutants since regular monitoring began in 1975. Unfortunately, this trend of improvement is not expected to continue in the future, as additional emissions reductions become more difficult and costly to achieve and as continued population and economic growth offset the effect of these reductions (SCAQMD, 1991).

THE TRANSPORTATION DISADVANTAGED

Although the private vehicle and the highway system provide unparalleled mobility, persons who do not drive or have access to private vehicles are greatly disadvantaged. Those without access to private vehicles are limited to using public transit or taxis, seeking rides from others, or walking or biking. Public transit provides limited service in most parts of the region, taxis are costly, and the region's dispersed land use pattern makes walking or biking a poor substitute for the car.

Persons who are transportation disadvantaged include people who are poor, older, or disabled and children. Vehicle ownership is related to household income: 60% of the region's carless households have annual incomes of $15,000 or less (Southern California Association of Governments, 1993). Lack of access to the private vehicle has the effect of reducing daily travel, both in person trips and vehicle driver trips, as Figure 13.3

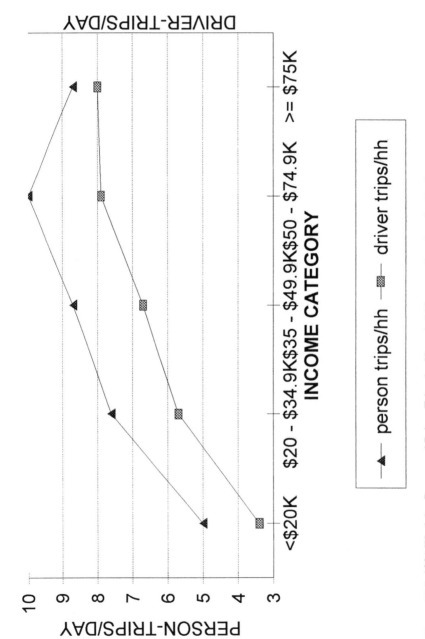

Figure 13.3. 1991 Weekday Person and Driver Trips, by Household Income, Los Angeles Region
SOURCE: Generated from data provided in Ochoa & Jones (1993).

238

Table 13.4 Travel Mode for Zero-Vehicle Households, by Household Income Category, in Share of Person Trips (Nationwide)

Income Category	Private Vehicle[a]	Public Transport	Walk	Other[b]
< $5,000	21.7	8.1	65.7	4.4
$5,000 - $9,999	41.1	18.3	36.6	4.0
$10,000 - $14,999	32.8	18.0	45.5	3.6
$15,000 - $19,999	37.7	18.8	37.2	6.3
$20,000 - $29,999	34.1	16.5	43.5	5.8
$30,000+	32.2	17.3	45.0	5.5

SOURCE: Lave & Crepeau (1994), p. 1.28.
a. Includes all private vehicle trips as driver or passenger.
b. Bicycle, motorcycle, and other.

illustrates. In the L.A. region, Los Angeles County has the largest proportion of zero-vehicle households (11%); Orange and Ventura Counties have the lowest proportion (5%).

How do people in carless households travel? Table 13.4 gives shares of travel mode by income category for zero-vehicle households on the basis of 1990 national survey data (comparable data for Los Angeles are not available). First, a larger proportion of trips are made by private vehicle than by public transit, meaning that borrowing cars and seeking rides from friends or family is a frequent mode of travel. Second, in all but one category, more trips are made by walking than by any other means. Third, for the lowest-income households, walk trips are apparently a substitute for transit trips, suggesting that for the very poor, transit fares are a significant barrier. These travel patterns suggest that public transit does not provide much mobility even for those who do not have private vehicles.

Policy Response to Transportation Problems

Reflecting concerns about air quality, energy, and the quality of life in this region, policymakers in Los Angeles rejected the strategy of increasing highway capacity in the 1980s and instead focused on providing alternatives to the automobile in an effort to curtail its use. These alternatives include a massive rail transit investment program as well as an array of transportation demand management (TDM) strategies designed to increase ride sharing and transit use.

THE RAIL TRANSIT PROGRAM

The L.A. region has made an enormous commitment to rail transit investment. This commitment is strongest in Los Angeles County, in which $78.3 billion of the 1993 30-year transportation investment plan was earmarked for rail transit capital projects (Los Angeles County Transportation Commission [LACTC], 1992). Policymakers hoped that an extensive rail transit system would reverse the trend of growing reliance on the automobile and, in the long run, redirect land use patterns to a more compact, higher-density urban form. By 1994, financial impacts of operating the first rail lines, together with severe recession and new federal guidelines that restrict capital investment plans to fund availability, forced a massive downsizing of the long-range plan. The 1995 plan totals $72 billion, of which $15.4 billion is budgeted for rail transit construction (Los Angeles County Metropolitan Transportation Authority [LACMTA], 1995).

Transit has continued to lose market share in Los Angeles as it has everywhere else, despite significant capital investment and operating subsidies. The Blue Line, the first rail line to open in Los Angeles, provides an example of the problems with rail transit investment. The Blue Line extends south 23 miles from downtown Los Angeles to the city of Long Beach, and its capital cost was nearly $1 billion. The 1992 Blue Line ridership was about 11.3 million. Recent passenger surveys show that about 30% were former automobile users; the rest were former bus riders or new travelers. The regular fare was a flat $1.10 (subsequently increased to $1.35 in 1995); these revenues cover just 15% of operating cost. These numbers suggest that the annual total subsidy cost (including both capital and operating costs) of each new rail rider is in excess of $20,000. The Metrolink commuter rail services, designed to attract long-distance commuters, are even more costly. The 1992 total subsidy cost per trip was $21.02, compared with $11.34 for the Blue Line and $1.17 for MTA buses (LACMTA, 1994).

Investment in rail transit is an expensive and inefficient way to divert trips from the automobile for several reasons. First, rail transit is costly to build and to operate. In contrast to the Blue Line, bus transit revenue in Los Angeles covers nearly 40% of operating costs, and some heavily used inner-city routes cover close to 90% (LACMTA, 1994). With the same subsidy dollar, far more public transportation can be provided with buses than with trains.

Second, although rail transit has much higher capacity, the dispersed travel patterns of Los Angeles make it impossible to take advantage of this capacity. In a region as extensive and decentralized as Los Angeles, even hundreds of miles of rail transit can provide reasonable access to only a small share of potential trip origins and destinations. The Los Angeles County rail transit plan is centered on downtown Los Angeles, and although the greater downtown area is still the single largest employment center in the region, it is losing market share as the decentralization process continues. The vast majority of both jobs and residences in the region will not be conveniently served by rail transit even if the entire planned system is built.

Third, there is no reason to expect the structure of Los Angeles to be substantially different even 30 or 40 years from now. High-density, compact cities developed before adoption of the automobile. In North America, compact cities were the result of the industrial revolution and the state of transportation technology. The technology of production required agglomeration, and access to the intercity rail network was critical. The streetcar system determined residential patterns for workers. None of these conditions exist today, and the continued shift to service and information-based activities suggests that economic activities will be even less "place dependent" in the future.

Finally, under these conditions, the only way to promote high-density development is through land use regulation, and in the United States, the ability of government to control land use is limited. A coordinated program of land use regulation would require cooperation among the nearly 200 municipalities in the region, would preempt the land use authority of these municipalities, and would require a historically unprecedented level of control over private property decisions.

Effective mass transit requires high densities, such as those of Manhattan or Tokyo, and the potential demand for such environments is uncertain (Downs, 1992). Moreover, even if higher densities were accomplished along rail transit corridors, these changes would have little overall impact on the region's travel patterns.

A study sponsored by the Southern California Association of Governments is instructive here (Urban Innovations Group, 1993). The study used a transportation forecasting model to determine the effect of various land use scenarios on transit use. Results showed that by relocating 75% of all forecast employment growth and 65% of population growth in the region to transit station capture areas and local activity centers, the work trip transit mode share would be in the range of 7% to 10%, compared with the current share of 5.6%. Study authors conclude that even if the anticipated land use changes were to occur, travel patterns would not change much because the regional patterns of land use would not change much.

TRANSPORTATION DEMAND MANAGEMENT

The second major transportation policy thrust in Los Angeles is TDM, which encompasses a variety of strategies to encourage ride sharing and transit use. These include carpool matching, sponsored vanpools, preferential parking for carpools, transit and ride-sharing subsidies, telecommuting programs, bike lockers and showers, and so on. Voluntary TDM programs have been provided by many large employers since the 1970s. Most recently, however, employer-based TDM programs have been legislated by state and local laws.

The most ambitious TDM program is Rule 1501, implemented in 1988 by the SCAQMD as part of its plan to achieve compliance with air quality health standards.[2] It requires both public and private employers having 100 or more workers at any given site to complete and file a plan by which they intend to increase average vehicle ridership (AVR) to a specified level within 1 year. AVR may be roughly defined as the average ratio of workers to vehicles arriving at the site during morning peak hours. At present, there are about 6,200 employers subject to Rule 1501. They employ about 3.8 million workers—about 40% of the region's workforce.

Although each work site must submit a plan for achieving the AVR target within 1 year, actually achieving the target is not mandatory. Thus, if a work site does not meet the target within 1 year, it must file another plan for meeting the target within the next year. The specified AVR targets, which vary by location, imply an average increase of 25% over current ambient levels, and thus require substantial shifts in mode to be achieved.

Recent studies of Rule 1501 show that it has had a small but significant impact (Giuliano, Hwang, & Wachs, 1993). After 1 full year of implementation, AVR at a sample of 1,110 sites increased by 3.4%, from 1.22 to 1.25. After 2 years of implementation, AVR at a sample of 243 sites continued to rise to 1.30. These AVR increases resulted from a significant increase in carpooling (and a significant decrease in driving alone). Vanpooling also increased but contributed little to AVR change. There were no changes in transit use, walking, bicycling, or telecommuting.

The purpose of Rule 1501 is to reduce auto emissions by reducing peak period travel, which is usually measured in total vehicle miles of travel. Because individual employee information is not available, VMT change can be only roughly estimated. Martin Wachs and I estimated a reduction of 325 million annual VMT, using the average regional commute trip length of 10 miles and calculating the number of peak period automobile trips reduced as a result of the observed increase in ride sharing. This amount is 0.4% of annual VMT in the region. This estimate is actually an optimistic upper bound of the impact of Rule 1501, because we have not taken into account the new trips that would be attracted as a result of more favorable traffic conditions. Given the level and extent of congestion within the region, it seems reasonable to expect that any reduction in peak period work trips would be offset by increases in other types of trips.

An assessment of the effectiveness of this TDM policy requires a comparison of program outcomes with program costs. Employers incur costs to provide ride-sharing incentives as well as to meet the procedural requirements of the regulation. Employers devise their own ride-sharing plan and can use any combination of incentives and disincentives (such as parking fees for solo drivers) to achieve the AVR target. Employers have overwhelmingly chosen to use incentives rather than disincentives, which, all else equal, translates to higher program costs. Cost estimates vary widely: They range from $12 to $263 per peak employee per year, and the annual cost to employers of removing one vehicle work trip has been estimated at around $3,000 (Ernst & Young, 1992). Although this amount is far less than the cost of vehicle trips removed by the Blue Line, it represents a significant additional cost of doing business for large firms.

Furthermore, it is unlikely that the AVR (trip reduction) targets will be achieved, unless employers are willing to offer large subsidies to alternative modes or impose large additional costs on driving alone. Such actions are unlikely, given the growing public opposition to the regulation. As a result of this opposition, Rule 1501 was temporarily suspended in 1994, and a special task force was established to propose alternative strategies for meeting the region's required trip reduction goals.[3]

EQUITY CONSIDERATIONS

The best available evidence in Southern California suggests that efforts to remove automobile trips from the road and divert them either to rail transit or, through TDM, to carpools, vanpools, or buses, as well as trains, can produce only modest results. The diversions that are accomplished are too small to have any significant impact on either congestion or air quality, and they result in high costs to private employers, public agencies, and the public. They also have negative impacts on those who are most dependent on public transit.

First, in part as a result of the high subsidy cost of rail system expansion, budget shortfalls have been met by reducing bus service, generating the ironic outcome of reducing total transit ridership and further reducing accessibility for those who are transit dependent. Second, rail commuters have higher than average household income, whereas bus commuters have lower than average incomes. Affluent rail commuters are subsidized at a much higher rate than are bus commuters under existing fare policies. As the rail system expands, these inequities will grow.

Third, the regional transit authority spends more annually on security for the rail lines than for the bus system in an effort to attract new riders to the rail services by assuring them that transit is safe. On a per passenger basis, for example, security costs are $1.29 for the Blue Line and $.03 for bus (LACMTA, 1994). Again, public funds are used to benefit more affluent, "choice" riders at the expense of less affluent bus riders. Bus crime victims are likely to be older and female and have low incomes (Levine & Wachs, 1986). For these people, fear of crime imposes another barrier to personal mobility. Finally, the single largest source of funds for the transit system is a county sales tax, a regressive tax that falls disproportionately on poor households.

Alternative Policies for Los Angeles

New policies must be established to solve congestion and pollution problems and to provide better transit service for those who need it most. To understand the difficulties involved in attempting to develop and implement such policies, it is necessary to consider the larger context in which transportation policy decisions are made.

At the heart of the difficulty of reducing automobile use is its heavy subsidy. Some examples include the exclusion of roads from local tax rolls, the support of traffic police and emergency services by property and other taxes, the provision of "free parking" to employees, and the failure to impose fees on auto use that reflect costs of the air pollution such use generates. As long as these subsidies exist, it will not be possible to entice significant numbers of travelers to what they perceive to be inferior modes, nor will it be possible to provide an alternative that is not perceived to be inferior by most travelers. And it will be financially impossible to provide enough heavily subsidized transit service to adequately serve those who are transportation disadvantaged.

It is important to understand that these subsidies are not a historical accident that can easily be reversed. They are the result of deliberate policy choices, and they reflect public preferences that strongly favor the private auto. I noted earlier that the automobile was adopted rapidly. Once the automobile revolutionized personal travel and trucks revolutionized freight transport, there was a broad base of political support for building and maintaining the appropriate infrastructure and for making the purchase and use of these vehicles as economical (to the user) as possible. As early as the 1920s, several states had established earmarked vehicle fuel taxes to support highway building and maintenance. The Interstate Highway Act of 1956, which funded the interstate highway system, is often viewed as a watershed decision in transportation policy but is better understood as a result of trends that had been in process for decades. The Highway Trust Fund established by the 1956 act remained largely intact until 1991. In addition, automobile fuel taxes, registration fees, and other taxes continue to be the lowest in the world, making auto ownership and use in the United States a financial bargain. These favorable conditions resulted in a predictable outcome: The United States has the highest rate of auto ownership and use of any country in the world.

It is precisely this strong support for the highway system that has led policymakers to choose the costly and ineffective policies just discussed. Thus, implementation of meaningful policy changes will require a fundamental change in attitudes toward the private automobile.

PRICE THE AUTOMOBILE TO
MORE CLOSELY REFLECT SOCIAL COST

The most effective way to reduce use of the private auto is to price it more appropriately. Some alternatives include a substantial increase in fuel taxes, vehicle registration fees based on mileage and fuel economy, elimination of parking subsidies, and even road pricing. Increasing the price of using the auto would increase demand for alternative transport modes, which would in turn stimulate the supply of alternative services (thus improving service quality) and reduce the subsidies required to operate them. Various pricing alternatives are being explored in response to the disappointing results of the policies described earlier. They have so far, however, received little public support.

Pricing policies are usually opposed on the basis of equity; lower-income travelers would be disproportionately affected. The inequities of pricing policies can be offset by the revenues they generate. For example, regressive taxes such as the sales tax that is currently assessed to provide transit capital funding could be replaced by automobile fee revenues. In addition, pricing policies can be structured to minimize their impact on low-income groups by establishing graduated fee structures or rebating fees to low-income households.

PROVIDE A WIDER RANGE
OF MASS TRANSPORT CHOICES

If mass transport is broadly defined as any alternative to the single-occupant auto, there are many opportunities for more effective transport services. Effective mass transport must be able to serve the dispersed spatial distribution of activity patterns, meaning that flexible modes such as jitneys, shared-ride taxis, and employer-based buspools, vanpools, and carpools are more likely to be able to compete with the auto than conventional fixed-route transit. These alternatives require far less subsidization than mass transit, and, as automobile subsidies are reduced, a wider range of such alternatives become economically feasible. A more diverse mix of mass transport choices, addressing the needs of specific market segments, is more likely to increase overall use of mass transit and contribute to congestion and air quality goals.

Barriers to providing these types of transit services are primarily regulatory. Airport shuttle services are prohibited from serving other types of destinations. Jitney services as well as shared-ride taxis are prohibited in Los Angeles and most other U.S. cities. These restrictions are mainly the result of opposition from public transit agencies and other vested interests that fear the competition for passengers that such services might generate. In a major policy shift, the 1995 LACMTA plan proposes a "mobility allowance" for areas with large numbers of transit users. The LACMTA would distribute operating funds directly to local jurisdictions, thus allowing MTA service to be replaced with shuttles or other flexible modes. If successful, the mobility allowance would make a wider range of transit choices possible.

TAKE ADVANTAGE OF NEW TECHNOLOGY

As stated earlier, technology has been the primary means by which automotive emissions have been reduced. Technology has also been the primary means for managing congestion. Examples include ramp metering on expressways, coordinated signal timing, and advanced incident surveillance techniques.

Rapid advances in communications technology are making possible dramatic improvements in both traffic management and mass transportation. There is no doubt that technology will be exploited to accommodate future automobile travel demand, and there is great optimism within the engineering community that the future is one of automated highways. New technology, however, holds promise for improving transit services in the shorter term as well. Using telephones or computer terminals, travelers could be informed of the location of the bus on a given route and its expected arrival at a given bus stop. Computers at employment centers could provide real-time carpool or vanpool matches. A "universal fare medium," or "smart card," used like a credit card, could be used to pay fares and transfers between different modes and could be linked to other transportation services such as taxis and parking fees. Some of these concepts are already in use; others should be available soon on a limited scale.

THE LAND USE ISSUE

Many urbanists and environmentalists believe that the fundamental problem in Los Angeles and other U.S. metropolitan areas is the existing land use pattern. They argue that as long as Los Angeles retains its low-density, decentralized structure, dependence on the auto and its attendant congestion and environmental problems cannot be reduced. Thus, they propose policies to promote high-density development, particularly around rail transit nodes.

I have already explained that attempts to significantly change urban form are unlikely to be successful, and, taken alone, would have little effect on the region's congestion and pollution problems. This does not mean, however, that such policies should not be pursued. To the extent that policymakers are intent on investing in rail transit, they have a responsibility to promote as much development as possible in rail corridors to encourage transit use.

Nor should high-density development be prohibited in other areas where it can be accommodated. Density restrictions often reflect public opposition to the provision of affordable housing or more generalized antigrowth sentiments. There may well be demand for mixed-use, pedestrian- and transit-oriented environments; planners and developers should have the opportunity to design and build them.

Finally, it bears emphasizing that high densities do not reduce traffic congestion. In fact, the opposite is true, as the images of Tokyo, Hong Kong, and Manhattan illustrate. Although these cities have far fewer auto trips per capita and people travel shorter distances than is the case in Los Angeles, the dense concentration of these trips results in more congestion.

Conclusion

Mobility is not a social, environmental, or economic problem. Rather, mobility provides access to opportunities for employment, health care, recreation, and social interaction. Improved mobility has increased access to these opportunities, and thus it may be argued that the goal of transportation policy should be to continue to increase these opportunities, rather than to restrict them (Webber, 1992). The challenge is to accomplish this goal while avoiding the negative consequences of congestion, air pollution, and inefficient energy use. The complexity of metropolitan structure, of individual decision making about residential and work location, and of U.S. politics makes it difficult to conceive of a single policy or technology that can promise to meet this challenge.

In this chapter, I have argued that efforts to reduce auto use by investing in rail transit or relying on TDM are misguided. Rather, auto users must be assessed the social costs that they are imposing on the urban system, and doing so will have a positive effect on both the environment and quality of life. If auto users are charged on the basis of social costs, a variety of mass transportation alternatives become both feasible and efficient. If auto users are not charged these costs, neither large investments in high-capacity transit nor extensive TDM programs can significantly affect modal choice.

Notes

1. Forecasts are from Southern California Association of Governments (1994) transportation forecast model estimates, based on 2015 regional population of 21.4 million and capital projects limited to those for which funding has been identified.

2. The rule was originally known as Regulation XV.

3. In December 1995, the SCAQMD made employee trip reduction programs voluntary. Employers have the option of paying a $61 per employee fee, continuing trip reduction programs, or participating in other emissions reduction programs.

References

Altshuler, A., Womack, J., & Pucher, J. (1979). *The urban transportation system: Politics and policy innovation.* Cambridge: MIT Press.

Calvert, J., Heywood, J., Sawyer, R., & Seinfeld, J. (1993). Achieving acceptable air quality: Some reflections on controlling vehicle emissions. *Science, 261,* 37-45.

Downs, A. (1992). *Stuck in traffic: Coping with peak hour congestion.* Washington, DC: Brookings Institution.

Ernst & Young. (1992). *South Coast Air Quality Management District Regulation XV cost survey.* Los Angeles: Author.

Environmental Protection Agency. (1993). *National air pollution emission trends: 1990-1992* (Rep. EPA-454/R-93-032). Research Triangle Park, NC: Office of Air Quality Planning and Standards.

Federal Highway Administration. (1990). *Highway statistics 1990.* Washington, DC: Government Printing Office.

Giuliano, G., Hwang, K., & Wachs, M. (1993). Mandatory trip reduction in Southern California: First year results. *Transportation Research A, 27*(2), 125-138.

Hanks, J., & Lomax, T. (1991). Roadway congestion in major urban areas: 1982-1988. *Transportation Research Record, 1305,* 177-189.

Hu, P., & Young, J. (1992). *Summary of travel trends: 1990 nationwide personal transportation survey* (Rep. FHWA-PL-92-027). Washington, DC: Federal Highway Administration.

Lave, C., & Crepeau, R. (1994). Travel by households without vehicles. *Travel Mode Special Reports* (1990 NPTS Rep. Series, pp. 1.1-1.47). Washington, DC: Office of Highway Information Management.

Levine, N., & Wachs, M. (1986). Bus crime in Los Angeles: II. Victims and public impact. *Transportation Research A, 20A*(4), 285-293.

Los Angeles County Metropolitan Transportation Authority (LACMTA). (1994). *A look at the Los Angeles County Metropolitan Transportation Authority.* Los Angeles: Author.

Los Angeles County Metropolitan Transportation Authority (LACMTA). (1995). *A plan for Los Angeles County transportation for the 21st century.* Los Angeles: Author.

Los Angeles County Transportation Commission (LACTC). (1992). *Proposed thirty year integrated transportation plan.* Los Angeles: Author.

Ochoa, D., & Jones, L. (1993). *1991 statewide travel survey final report.* Sacramento: State of California Department of Transportation, Office of Traffic Improvement.

Rosetti, M., & Eversole, B. (1993). Journey to work trends in the United States and its major metropolitan areas, 1960-1990. Washington, DC: Office of Highway Information Management.

Small, K., & Kazimi, C. (1994). *On the costs of air pollution from motor vehicles* (Irvine Economic Paper No. 94-95-3). Irvine: University of California, Department of Economics.

South Coast Air Quality Management District (SCAQMD). (1991). *1991 air quality management plan.* Los Angeles: Author.

Southern California Association of Governments. (1993). *1991 Southern California origin-destination survey summary findings.* Los Angeles: Author.

Southern California Association of Governments. (1994). *Final regional mobility element.* Los Angeles: Author.

Urban Innovations Group. (1993). *Regional urban form study.* Report prepared for Southern California Association of Governments, Los Angeles.

Vincent, M., Keyes, M., & Reed, M. (1994). *NPTS urban travel patterns: 1990 nationwide personal transportation survey* (Rep. FHWA-PL-94-018). Washington, DC: Office of Highway Information Management.

Webber, M. (1992). The joys of automobility. In M. Wachs & M. Crawford (Eds.), *The car and the city: The automobile, the built environment, and daily life.* Ann Arbor: University of Michigan Press.

"Can't We All Get Along?"

THE REVEREND CECIL MURRAY
First AME Church, Los Angeles

Our idea of nothing to fight about is fighting about complexion. The complexity of complexion. What difference does it possibly make what color or colorlessness you are? My idea of nothing to fight about is fighting about the shape of one's nose or lips or hair or heel or derriere, as compared to the shape of one's character. So the starting answer is obviously *no,* we can't get along. We have never gotten along. At least the record indicates we have never gotten along.

We have never *had* to get along, up to now. Now we have a different set of equations. Now we have a different Calculus of Probabilities. What makes the difference? That hole in the ozone, that community call to concern. Necessity is the mother of invention. The children of necessity now point up to that hole and say, "*That is calling us to care for each other.*" And so we are forced to do the same thing that porcupines must do in the winter. They must huddle together even though their quills offend each other. They must huddle together and suffer together as community or perish apart as fools. So now we can get along because we *must* get along. Now the "haves" and the "have nots" *must* get along.

America does have a hole in its soul. Los Angeles does have a hole in its soul—140 different nations and we're trying to see who's going to be on the top and who's going to be on the bottom, rather than trying to see who can come up with the way that we can all peacefully coexist. We don't have to like each other, but we've got to love each other. We're going to offend each other with these quills as we porcupines huddle together. But since no one wants to die, we'd better learn how we can live.

I think a starting need is at the presidential level or the governmental level of the state, or the city, or the campus. (Perhaps USC will give it to us.) We need a single agenda. We need a national agenda. We need a regional agenda. We need a city agenda, an agenda with a wish list. What are we asking the educational world for? What are we asking of the corporate world, the religious world? What are we asking of the grassroots world? What is our wish list for this master agenda? What are its time parameters: short-term, midterm, and long-term?

A second item that might help to fill this hole in our soul and help us get along is to take the military bases that are already our property, governmental property, as they downsize and de-escalate. Let's take these and use them to revamp human talents. Let's take these and use them to equip the unequipped. We have 35,000 homeless people in Los Angeles alone, 180,000 homeless in the county. We have a 1 in 4 chance of a Black child dropping out of school before graduation and thus becoming a drain on the economy. We have people who are unskilled and underskilled going into the space age of the 21st century. We can't just sit fat, dumb, and happy and pretend that they are not going to be a drain. Already the health services are drained to the breaking point. And we can't just throw money at our problems. We've got the governmental bases. We have the American know-how. We have the American Will. Seventy-five percent of any budget is payroll. Then perhaps we can volunteer that 75%. We can volunteer woman-hours and man-hours for a month. We can say, "I commit to this." We will drive the 20 miles to the base, and there with our skills, we will recycle the unskilled. That will help us fill that hole in our soul and in our economy.

A third way is entrepreneurial development. It isn't enough to offer a job to the minorities in this city. We must find a way also to make them jobbers. We must get a way of giving microloans, small loans. The heart of American enterprise, 80%, is small business. The conglomerates are the unusual thing, and even they are in jeopardy. Right now, America is being carried by the small-business development. On our local level, with the modest gift of $1 million by the Disney Corporation—portions from their employees, portions from management—we have encouraged development of small businesses. We have taken that $1 million and leveraged it to $6 million. We have started up 35 new businesses. We have renovated 35 existing businesses. We will ultimately hire 500 persons. But it isn't enough to give a $20,000 loan. We have a prior 10-week period in which we indoctrinate them. We have professionals in the field who give pro bono time to lecture. Then we give a loan. Then we hand-hold them for 2 years with MBAs, CPAs, and attorneys from the church, USC, UCLA, Southwest College, and Disney. They are willing to give this time. Results? With the business failure rate some 40% in America, some 65% in Black America, we have had a 5% failure rate. All have done well with the hand holding and monitoring.

A fourth concept has to do with a national plan for *mentoring,* one that can be implemented even on the block level. Every youth aged 7 to 17 needs a mentor. And when a young boy or young girl reaches 18 years of age, that person needs to reach back and take a mentee, committing to so many hours of visitation for a given period: phone calls, visits, modeling—the extended-family concept.

We've only broached the surface of a complex challenge. I know that my mouth is no prayer book and there are many difficult questions to answer, but we must address the most essential question of all: Can we get along? The universe and all of its creatures respond: "Can we get along? We'd better!"

Re-Imagining
Los Angeles

14

ROBERT FISHMAN

"Unreal city, full of dreams"—so Baudelaire described Paris in the mid-19th century (quoted in Benjamin, 1983, p. 18), and so might Los Angeles be described in the 20th century. For Paris, the dream was to be, as Walter Benjamin put it, the "Capital of the Nineteenth Century" (p. 5): the central place where power, wealth, and culture met. Los Angeles in our time has imagined itself as a different type of capital, the capital of the 20th-century Good Life, the anti-Paris, or more precisely, the anti-New York. In place of dense, vertical agglomerations of people and power, there would be a great region open for movement and enjoyment. No skyscrapers, but no slums either; instead, a garden landscape assembled from multitudes of single-family houses would stretch in every direction. No great public squares or urban monuments, but the city would embody the new freedom created by fast cars hurtling along wide freeways from snow-capped mountains to Pacific beaches.

From our vantage point at the end of the 20th century, we can see that the Los Angeles "city of the future" has no future. The dream of limitless mobility has been swallowed up in endless sprawl and nonstop congestion. Now among the densest of American cities, Los Angeles has generated its own form of hierarchical constriction and its own form of slums. To be sure, new developments at the edge of the region still push out into the desert in a naive search for open spaces and uncrowded freeways. But they are like the provincial acolytes of a dying religion who haven't heard that the temples of the metropolis are already closed.

How, then, to re-imagine Los Angeles when the very idea of the "city of the future" has itself been discredited? I suggest that the key lies not in a more extravagant science fiction rendering of the city of limitless mobility. Still less can the re-imagined city be found in a passive acceptance of the Los Angeles of today, rebaptized as "the postmodern metropolis" and all its sins forgiven.

For me, the way to re-imagine the future of Los Angeles lies paradoxically in an imaginative reconstruction of its past. Before the freeway-and-sprawl Los Angeles took shape in the mid-20th century, there was another city, built up on the armature of perhaps the best mass transit system of any region in the world, a city that embodied both human-scale communities and a healthy balance between nature and the built environment. This "other Los Angeles" still exists, buried as it were beneath the sprawl-city, but still available as a vital alternative form for future planning.

In the 1920s, the Los Angeles region possessed three great attributes that speak powerfully to us today. First, there was a balanced transportation system in which the emerging highway system coexisted with the thousand-mile Pacific Electric trolley system. Second, decentralized growth was still disciplined into well-defined towns and villages laid out along the main Pacific Electric routes. Third, these routes converged to maintain the downtown as a lively, open, and diverse regional center. To be sure, this was a region of 1 million people, compared with 14 million in 1995. Nevertheless, I believe it is worth asking: What were the basic principles of Los Angeles urbanism of the 1920s, and what relevance do these principles have for re-imagining the city today?

Reyner Banham (1971, Chap. 4) was perhaps the first to note that the decentralized form of 20th-century Los Angeles was shaped first by the streetcar system, not by the automobile. In Los Angeles, as elsewhere throughout the United States and Europe, the streetcar systems introduced between 1895 and 1925 a wonderfully urbane and humane technology characterized simultaneously by freedom and discipline. Before electric traction, the horsecars and steam railroads were restricted by cost to the middle class, who used them to escape the crowded cities and pursue their suburban dreams. But the technology of electric traction was the first to provide speedy service that a majority of the population could afford. Riding the trolley, even working-class urbanites could escape the crowded slum-tenement-and-sweatshop districts of the urban cores. The great nightmare of 19th-century urbanism—the implosion of population into ever denser slum "jungles"—was overcome.

Electric traction thus brought freedom to escape the confines of the urban core, but compared with the later automobile, it also brought a real discipline. Population could spread out only within walking distance of transit stops. This encouraged the formation of coherent local neighborhood centers, leaving the land beyond walking distance still undeveloped. Moreover, these lines worked best in a hub-and-spoke formation, with the regional downtown constituting the hub in which regional institutions necessarily clustered. The logic of the streetcar system thus created a dynamic symbiosis between downtown and the neighborhoods (see, e.g., Warner, 1962).

These features of streetcar urbanism were common to all large cities throughout the world at the turn of the century. But I believe only Los Angeles was able to capture the truly revolutionary potential of the streetcar region. In the cities of the U.S. East Coast and Midwest (as well as European cities), the solid mass of traditional urban form inhibited decentralization. Congestion at the core meant that lines had to be constructed either as subways or elevateds, increasing costs and limiting outward movement. Moreover, the already dense factory zones of the great cities tended to maintain their functional concentration. Densities remained high even in the electric traction era, with multifamily housing or closely packed single-family dwellings still the rule for the working class and lower middle class.

Los Angeles, however, experienced the full capacity of the streetcar system to decentralize population while maintaining a vital regional hub. The extraordinary personal ambition of Henry Huntington, founder of the Pacific Electric system, exploited the opportunities presented by a lightly settled but booming region to push the technology of electric traction to its limits. Like the Eastern traction magnates Harry Elkins and George Widener, Huntington tied streetcar service to real estate speculation; each new line or extension meant land that could be profitably developed along its corridor. But Huntington operated in a different environment from his Eastern counterparts. Mass transit systems in the East and Midwest cut through dense immigrant neighborhoods, permitting some expansion at the edge of the built-up urban districts. Huntington dramatically flung his lines out from central Los Angeles, crossing open countryside to connect the downtown to the Pacific Ocean, the San Fernando Valley, and the San Gabriel foothills (Fogelson, 1993, Chap. 8).

Although the Eastern traction magnates served a poor immigrant population tied to dense factory zones, Huntington's Pacific Electric served the relatively affluent migrants to Los Angeles who were never tethered to the small factory zone that was developing east of the downtown but who were free to follow the opportunities for good living and profitable small businesses throughout the region. These migrants sought out the scenic towns and villages spread along the trolley lines.

The Los Angeles that was emerging in the 1910s and 1920s thus bore a surprising resemblance to the planning theories that were emerging during those years from the Garden City movement in England and the Regional Planning Association of America (RPAA) in the United States. In 1928, Clarence Stein, Henry Wright, Lewis Mumford, and other members of the RPAA founded the "New Town" of Radburn in Bergen County, New Jersey, as the model for a New York region of the future based on decentralized settlements (Schaffer, 1982). But Los Angeles already had scores of spontaneous New Towns. Such local centers as Hollywood, Pasadena, and Santa Monica were in effect Garden Cities, coherent settlements still surrounded by a greenbelt of irrigated farms and citrus groves. These New Towns, to use the phrase that Garden City advocates had just adopted, had their own local centers and local character, but they were also joined efficiently to the regional downtown by Pacific Electric (Bigger & Kitchen, 1952).

Los Angeles in the 1920s promised to become a new type of urban region. Instead of a dense mass of factories and tenements surrounding downtown, the built-up area of the city soon gave way to a landscape of fields and orchards dotted with still separate towns. The scattered movie studios—soon to be joined by even more dispersed aircraft factories—foretokened an advanced industrial economy that was integrated into a still verdant landscape. Freed from smoke pollution, overcrowding, and slums, Los Angeles would become "the garden metropolis."[1]

This vision of a garden metropolis based on electric traction did not fall victim to the technological obsolescence of the streetcar system. Indeed, transit engineers in 1925 proposed a wonderfully intricate plan of short subway tunnels, elevated lines, and dedicated rights-of-way that would have kept the Los Angeles transit system running efficiently well into the second half of the century (Kelker, De Leuw, & Co., 1925). Still less, as Scott Bottles (1987) has effectively shown, was Pacific Electric "done in" by a highway conspiracy led by General Motors. Instead, as I have argued elsewhere, the discipline imposed by the streetcar system was felt in the boom years of the 1920s to be an intolerable limitation on real estate speculation.

In the great 1920s real estate boom, both humble curbstone speculators and large operators such as the Chandler family were seized by the mania of a rapid conversion of the vast Los Angeles basin into lots for single-family homes. The problem was that if the streetcar remained the dominant form of transportation, only lots close to the lines could be sold. The lines could, of course, be expanded, but the essential discipline of the streetcar system would remain. By contrast, a grid of new automobile roads could open up the whole county for immediate sale.

The crucial decision that I believe has shaped Los Angeles since the mid-1920s came when Angelenos were forced in 1924-1925 to choose between upgrading the Pacific Electric system (now out of Huntington's control and a quasi-public utility) and a major street traffic plan (Olmsted, Bartholomew, & Cheney, 1924) put forward originally under the auspices of the Automobile Club of Southern California. The automobile plan was the real estate speculator's dream: a vast grid of wide boulevards that would put any potential subdivision "inside the grid" and thus plugged into the automobile network of private transportation. By contrast, the Pacific Electric plan would have inevitably resulted in development limited to the light-rail corridors. This would have preserved the townscape of the Los Angeles region, both in the walking-scale vitality of the villages and in the greenbelts that surrounded them. It also would have kept the downtown as a lively and necessary urban core for the region. But the Pacific Electric plan would have sold fewer lots (Fishman, 1987, chap. 6).

Given the centrality of real estate speculation to the Los Angeles economy in general and to the Los Angeles elite in particular, the victory for the major traffic street plan was inevitable. Ironically, the boom of the mid-1920s soon cooled and gave way to the 1930s bust, which left almost one third of the built-up area of the city as vacant lots. But the street plan prevailed (the boulevards are now the network of surface roads that still carry the bulk of the city's traffic), and so too did its ultimate logic: the deterioration of the trolley system, the obsolescence of both the local centers and the downtown, and the swallowing up of open space in urban sprawl.

As Greg Hise (1993) has shown, the 1920s ideal of relatively coherent development lingered into the 1940s and even 1950s, especially in the New Towns that grew up around the aircraft factories, but the deeper pattern was one of relentless fragmentation, as large and small developers responded to any opportunity to build on the former farms and vacant lots at hundreds of points scattered throughout the region. Poorly served by a deteriorating mass transit system, the downtown decayed and its functions exploded over the landscape in strip developments and shopping centers. These automobile-oriented commercial strips in turn supplanted trolley-oriented local centers. To compensate for this intense fragmentation, Los Angeles adopted a supergrid of freeways that could connect all the scattered destinations into a high-speed regional system.

By the mid-1950s, Los Angeles had earned its reputation as the "shock city" of the second half of the 20th century. For a brief moment, the city seemed to express the definitive form of the American dream: universal home ownership, universal automobile use, and unlimited access to high-paying jobs and exciting leisure facilities throughout the region by high-speed automobile travel direct from one's home to one's destination.

But this "new city for the 20th century" has proved to be a cruel delusion for so many of its residents. One might argue that the very power of the Los Angeles vision proved to be its undoing. As the economy boomed, migrants—first from around the country and then from around the world—flowed in, each hoping for a piece of the dream. But by the 1960s, increased population density meant exploding housing prices, intractable congestion, and new forms of segregation. Instead of the traditional central city juxtaposition of "The Gold Coast and the Slum" traced by the Chicago school of sociology in the 1920s, one had the low-density ghetto of South Central Los Angeles, an area larger than Paris, isolated economically and socially from the region. Meanwhile, the wealthy ensconced themselves in scenically spectacular but environmentally precarious hillsides and canyons, and the middle class spread out ever more remotely in what Banham (1971) called "the Plains of Id" (p. 161). Even before 1992, when the most destructive urban riots in American history rocked South Central, the Los Angeles "autopia" was dead.

Of course, one might argue that the failure of autopia has removed us still further from the trolley era and that to attempt to revive any of the urbanistic values of that earlier period—not to mention the trolleys themselves—is to go against the force of history. We are, we are told, in a postmodern, postindustrial period of "de-centering" when we must embrace fragmentation and impermanence, what Rem Koolhaas (1995) has called "the death of urbanism":

> Redefined, urbanism [must] accept what exists. We were making sand castles. Now we swim in the sea that swept them away. We have to dare to be utterly uncritical; we have to swallow deeply and bestow forgiveness left and right. Since we are not responsible, we must become irresponsible. (p. 19)

Against this profoundly negative and profoundly tempting philosophy, there remains the stubborn sense that (however deeply we swallow) the waste, ugliness, disorder, and injustice that distort people's lives in a failing city are not so easily forgiven.

As the Los Angeles autopia reaches a state of permanent crisis, the seemingly buried Los Angeles of the era of trolleys, local centers, and a vital downtown has become increasingly "visible" as a vision of an alternative future. I do not think I exaggerate when I assert that every major "progressive" project put forward by the most committed avant-garde designers constitutes a revival of the best features of the 1920s city. This, for me, is not criticism but praise. To be sure, the farms and citrus groves and local centers of 70 years ago will never magically reappear. But the humane and rational elements of an earlier period of urbanism can still serve to broaden today's options.

The most obvious and important project in this context is the work of the Metropolitan Transit Authority to make Los Angeles in the 21st century once again a world center of light-rail transportation and also an important commuter rail and even subway city. Although the 1992 regional rail plan has already been scaled back, the impetus toward a balanced transportation system for Southern California seems unstoppable. The larger question remains whether the massive investment in rapid transit can revive a more rational and disciplined urban form both in the local centers and downtown. As both the critics and proponents of rapid transit point out, re-creating rail as a viable alternative depends ultimately on re-creating some semblance of the urban order that existed before the major traffic street plan. Not only does one need well-defined local centers to provide a sufficient population within walking distance of the stops, but also one needs a vital downtown hub to provide an attractive and important destination most easily reached by mass transit.

Several parallel developments and plans point in a positive direction. Perhaps the most significant of these is the revival of so many of the region's local centers, in advance of any mass transit links. In part, this revival reflects the bankruptcy of the older freeway ideal of unlimited rapid access to all parts of the region. Faced with chronic traffic congestion, people from Santa Monica to Pasadena are rediscovering the local. At the same time, the emergence of Los Angeles as the nation's most important immigrant destination has given the city what it lacked in the 1920s: an intense diversity of ethnic neighborhoods with highly active commercial cores.

These factors together have revived for Los Angeles something that critics thought had disappeared forever: vital public spaces. Exemplified by the revival of downtown Pasadena and the success of the Third Street Promenade in Santa Monica, these spaces are characterized by their modest yet perceptive use of the opportunities that survive from the fabric of older cityscapes. They are at once historic preservation and a progressive response to a felt need for the revival of public space and the public sphere in general.

Beyond revivals of earlier cityscapes, Los Angeles is now the site of perhaps the most ambitious project in the nation to re-create a "new urbanism" on the basis of past values: Playa Vista, on the 1,000-acre site of the old Hughes aircraft plant and airport just south of Marina del Rey. Designed by a team led by active proponents of the new urbanism, Elizabeth Moule, Stefanos Polyzoides, Andres Duany, and Elizabeth Plater-Zyberk, Playa Vista is a conscious attempt to re-create the values of the 1920s Los Angeles. The project embodies relatively high-density, mixed-use, low- and midrise buildings to create a walking community with generous landscaped public plazas. Housing is predominantly multifamily, with courtyard apartments, duplexes, and quadruplexes as well as town houses. Streets are designed to encourage walking and to discourage automobile use. Finally, Playa Vista has its own greenbelt, a remarkable ecological setting designed by the landscape architect Laurie Olin. More than half the site has been set aside for open space, most notably, a wetland preserve, riparian corridor, and open bluffs overlooking the site, thus recapturing some of the balance of built and open space that Los Angeles used to possess (Katz, 1993b).

In December 1995, DreamWorks SKG, the movie studio founded by Steven Spielberg, Jeffrey Katzenberg, and David Geffen, announced plans to build a major studio complex at Playa Vista. The studio would cover more than 100 acres in the part of the project already designated for office and research employment. The DreamWorks plan includes offices, 15 sound stages, and bungalows for actors, directors, and producers—all grouped around an 8-acre artificial lake. Just as the great movie studios of 1920s Los Angeles anchored the growth of the 1920s metropolis, so the DreamWorks complex in Playa Vista—the first studio "back lot" to be constructed in Los Angeles in more than 60 years—combines a bold reassertion of L.A.'s continued primacy in world communications with a commitment to the urbanistic principles that had dominated Los Angeles in its most creative era. In Playa Vista, the city's past truly becomes its future (Rainey, 1995; Sterngold, 1995).

Yet the creation and re-creation of these local centers puts back on the agenda the question of downtown Los Angeles. As I have argued, the source of the urban dynamism of the 1920s rail city was the interplay between downtown and the local centers. The potentially isolating local neighborhood (especially when it was defined in ethnic and racial terms) was compensated for by a downtown that served as a true center for all the peoples of the region. Conversely, the potentially alienating impersonality of downtown was balanced by the local neighborhood community.

Does Los Angeles in the 21st century need a downtown on the 1920s model? Can L.A.'s downtown become once again a true center for its region? We must begin by admitting that any genuine downtown revival must first overcome a 70-year legacy of neglect and then of ruthless interventions to create a corporate downtown that excludes the public spaces that the American downtowns of the past were built to embody. Not surprisingly, many of those most concerned with social justice have seen any downtown revival plan as nothing more than a more or less subtle form of displacement of poor and homeless persons who now dominate the street life of the area.

I argue, however, that to give up on L.A.'s downtown is to give up both on the prospects for an efficient mass transit system transforming the region and on the ideal of a regional common ground beyond the racial, ethnic, and economic turfs that presently divide the region. Perhaps the most important initiative that takes these ideals seriously is the Downtown Strategic Plan, prepared by Elizabeth Moule and Stefanos Polyzoides for the City of Los Angeles Community Development Agency, with the collaboration of many of those who were working on the Playa Vista project (Katz, 1993a). Unlike the downtown plans of the past, the strategic plan is not a corporate plan. It emphasizes instead a diverse mixture of uses and building types. Moreover, it is not a redevelopment plan in the sense of calling for extensive demolitions and massive new construction.

Rather, the plan consists of 16 "catalytic projects" put forward to jump-start revival, all taking account of the downtown's projected status as hub for a regional rail system. These projects range from cluster housing around St. Vibiana Cathedral, to designating a Broadway theater district organized around preserving the historic movie houses, to better integrating the new convention center into the fabric of downtown life. Perhaps the most typical is the one to create a "Market Square" to use the existing wholesale fish and vegetable markets to draw shoppers from around the region to a lively outdoor-indoor retail market. Already one can see that the strategic plan was overly optimistic in its projections for mass transit projects. But the plan's attempt to address community development, economic growth, and social equity simultaneously, as well as its use of relatively small-scale mixed-use projects, makes it a model for downtown development. Most important, in my view, the plan has used a profound and thorough understanding of the former role of the Los Angeles downtown as a way of re-imagining its future.

Great cities have always re-imagined themselves through a complex process of innovation and remembrance. The need to recapture the past is at least as important as the need to adjust for the future. When Pope Sixtus V in the late 16th century created the great avenues to connect the pilgrimage churches of Rome, he was both re-imagining Rome's place in Baroque Europe and reviving the heritage of ancient Roman classical town planning. Baron Haussmann's 19th-century rebuilding of Paris was designed both to adapt the city to the needs of industrial capitalism and to reaffirm the classical heritage of the French metropolis.

As Los Angeles struggles to cope with the crises of a postmodern, postindustrial, posturban world, the city requires both innovation and a renewed sense of what is valuable in its past. Precisely because the present is dominated by political, environmental, and social disintegration, we turn to the past—not out of nostalgia or escapism—but for those models of working democratic institutions and vibrant public spaces that we urgently need to build the future.

*   *   *

Richard Neutra (1941) told the story of attending the 1931 International Conference for Modern Building (CIAM) at which the leaders of modern architecture attempted to solve the problem of the modern city. They had commissioned a series of maps of major cities all done at the same scale; the map of Los Angeles dwarfed the others. "To the puzzled amazement of European students," Neutra recalled, "business zones, for example, seemed to stretch for hundreds of miles . . . [while] multi-storied slums . . . seemed anomalously absent." The Europeans wondered, "Was this metropolis a paradise, or did there exist here a type of blight which fitted none of its classical descriptions?" (p. 191).

More than 60 years later, the question still seems oddly pertinent. If, as the contributors to this volume have shown, Los Angeles is no paradise, the city nevertheless fits none of the paradigms of urban crisis inherited from the past. In particular, the puzzlement of the European CIAM architects underlies the problem with which the editors of this book began their preface: the problem of the supposed exceptionalism of Los Angeles.

As the editors point out, both the critics and the boosters of the city embraced—for different reasons—the banner of exceptionalism. For the critics, Los Angeles was a city that broke all the rules and thus had to be marginalized as a curious exception to the general pattern. For the boosters, Los Angeles was proudly proclaimed to be an anomaly, hence exempt from the poverty, slums, overcrowding, immigrant hordes, and labor unrest that plagued other cities. For both the critics and the boosters, Los Angeles was a city that in effect stood outside history, that is, outside the main line of urban development that had produced the centralized industrial metropolis exemplified by Chicago.

If there is a single lesson from this book, it is that Los Angeles is now no longer an exception. What had appeared in the 1930s as anomalies were in fact the harbingers of the decentralized, polycentric development that would overtake all the great cities of the world. Far from standing outside urban history, Los Angeles has come to exemplify the great city of the late 20th century.

This is decidedly a mixed blessing—as this volume teaches. If Los Angeles has risen to the grandeur of a global city, its problems now define the urban crisis of our own time. Even the imagery of the crisis now reflects L.A.'s leadership. Urban pollution is now pictured as the smoggy haze over the Los Angeles basin rather than the smoky pall of innumerable smokestacks; urban congestion is a freeway at rush hour instead of packed subway cars; urban poverty now means South Los Angeles instead of the crowded tenements of New York's Lower East Side.

Despite the new imagery, the underlying problems are hardly new to great cities. Chicago in its prime at the beginning of the 20th century was exactly what the editors of this volume call Los Angeles today: "a 'First World' city flourishing atop a 'Third World' city." Chicago in 1900 was receiving an influx of poor immigrants whose diverse and conflicting backgrounds were unprecedented in world history; the city suffered from industrial pollution that threatened simultaneously to poison the air and to pollute its water sources in Lake Michigan. The city's inhabitants suffered from organized crime and municipal corruption. Meanwhile, frequent recessions produced a demoralized underclass as well as violent labor unrest that many observers felt was the prelude to inevitable social revolution.

But Chicago in its prime also possessed a coherent metropolitan form, the product of rail-based technologies that privileged the urban core. Its expanding industrial base was built on a productive and stable relationship to its region and to the national economy. The city was led by a coherent elite that strongly identified the city's well-being with its own. Chicago also boasted a powerful political machine that connected even the slum dwellers to the political process, that mediated among the conflicting ethnic groups, and that provided increasing services to the voters. Finally, despite deep class divisions, Chicago was held together by an ethic of assimilation that saw even the newest immigrants and worst poverty yielding to constructive citizenship.

Los Angeles today can boast far greater resources than turn-of-the-century Chicago: It is, on the whole, preferable to be entertainer-to-the-world instead of hog-butcher-to-the-world. But, as the contributors to this volume show, Los Angeles suffers from fundamental disjunctions that seem (from our perspective at least) to be far more disturbing than the problems of Chicago and the cities of its era.

Whereas the Chicago elite possessed a clear blueprint for good urban form as summed up in the famous "Burnham Plan" of 1909, Los Angeles must contemplate vast projects in an extended region whose form defies rational analysis or political control. Even if we possessed a coherent vision for such a region (which we do not), there is no coherent elite to carry it out. One cannot define L.A.'s long-term economic interests; instead, the region seems open to a quickly changing global economy that interacts in unpredictable ways with local conditions. Still less can one predict with any confidence the assimilation of migrants and immigrants when the underlying structure of the economy seems itself so unstable and unpredictable.

As this volume shows, politics in Los Angeles is as fragmented and decentered as the society itself. "The postmodern metropolis is increasingly polarized along class, income, racial, and ethnic lines," the editors assert in Chapter 1. Meanwhile, a postmodernist politics leads directly to the division of the city into isolated fiefdoms in which the old political machine's equation of votes-for-services no longer has any meaning. Although Chicago undertook remarkable public projects—from building schools and hospitals and libraries to reversing the flow of the Chicago River to fight pollution—Los Angeles seems caught in paralysis and privatization. Not only does this paralysis benefit only the rich who can afford privatization, but the public sphere is hollowed out precisely when vigorous action seems most necessary.

It is the grandeur and the misery of Los Angeles that these interlocking crises are not exceptional; they are to different degrees the fate of all great cities in our time. Thus, "rethinking Los Angeles" necessarily goes beyond the future of Los Angeles and of Southern California. It means ultimately confronting the fate of urbanity in the next century. It means reconceiving the 21st-century city as a human environment.

Note

1. See, for example, the eccentric, but to me revealing, pamphlet by Hunter (1923, p. 4).

References

Banham, R. (1971). *Los Angeles: The architecture of four ecologies.* Harmondsworth, UK: Penguin.
Benjamin, W. (1983). *Charles Baudelaire: A lyric poet in the era of high capitalism* (H. Zohn, Trans.). London: Verso.
Bigger, R., & Kitchen, J. (1952). *How the cities grew: A century of municipal independence and expansion in metropolitan Los Angeles.* Los Angeles: Bureau of Governmental Research.
Bottles, S. L. (1987). *Los Angeles and the automobile: The making of the modern city.* Berkeley: University of California Press.
Fishman, R. (1987). *Bourgeois utopias: The rise and fall of suburbia.* New York: Basic Books.

Fogelson, R. (1993). *The fragmented metropolis: Los Angeles, 1850-1930.* Berkeley: University of California Press.

Hise, G. (1993). Home building and industrial decentralization in Los Angeles: The roots of the postwar urban region. *Journal of Urban History, 19,* 95-125.

Hunter, S. (1923). *Why Los Angeles will become the world's greatest city* [Pamphlet]. Los Angeles: H. J. Mallen.

Katz, P. (1993a). Downtown Los Angeles strategic plan. In P. Katz, *The new urbanism: Toward an architecture of community* (pp. 212-219). New York: McGraw-Hill.

Katz, P. (1993b). Playa Vista. In P. Katz, *The new urbanism: Toward an architecture of community* (pp. 178-191). New York: McGraw-Hill.

Kelker, De Leuw, & Co. (1925). *Report and recommendations on a comprehensive rapid transit plan for the city and county of Los Angeles.* Chicago: Author.

Koolhaas, R. (1995, Winter/Spring). Whatever happened to urbanism? *GSD News,* 19. (Available from the Harvard Graduate School of Design, Cambridge, MA)

Neutra, R. J. (1941). Homes and housing. In G. W. Robbins & L. D. Tilton (Eds.), *Los Angeles: Preface to a master plan* (pp. 191-201). Los Angeles: Pacific Southwest Academy.

Olmsted, F. L., Jr., Bartholomew, H., & Cheney, C. H. (1924). *A major traffic street plan for Los Angeles.* Los Angeles: Traffic Commission.

Rainey, J. (1995, December 5). DreamWorks picks L.A. site for studio. *Los Angeles Times,* p. A1.

Schaffer, D. (1982). *Garden cities for America: The Radburn experience.* Philadelphia: Temple University Press.

Sterngold, J. (1995, December 14). Vast new DreamWorks film lot. *New York Times,* p. C11.

Warner, S. B., Jr. (1962). *Streetcar suburbs: The process of growth in Boston 1870-1900.* Cambridge, MA: Harvard University Press.

Name Index

Alatorre, Richard, 159
Alibrandi, Joseph, 209 (n 5)
Anaspach, R. R., 87
Annenberg, Walter, 209 (n 3)

Bakeer, Donald, 139
Bartlett, Dana, 5-6
Banfield, E. C., 61
Banham, Reyner, 2-3, 252, 255
Bartholomew, Harland, 7
Baudrillard, Jean, 69, 135
Benjamin, Walter, 251
Bollens, John C., 58, 64-65
Bottles, Scott, 254
Bowron, Fletcher, 59
Boyar, Louis, 49
Bradley, Tom, 68, 73 (n 5), 84-85, 156, 157, 160, 162, 214
Brah, A., 183
Braudy, Leo, 113
Brecht, Bertolt, ix
Brown, Elaine, 139, 142, 214
Brown, Scott, 22
Bunnett, Robert, 110
Burke, Yvonne Braithwaite, 159
Burnett, Charles, 134
Bush, George, 156

Cain, B. E., 84
Callicott, Ransom M., 57
Calvino, Italo, 3
Cannon, Shiela, 174
Chandler, Harry, 66
Chang, Edward, 158

Chavis, Ben, 177
Choi, Brian, 166
Choi, Cindy, 163, 164
Clinton, George, 143, 144
Cohn, Carl, 105
Coltrane, John, 140
Comstock, Sarah, 5
Cooley, Joe, 138
Cosby, Bill, 127
Cousineau, M. R., 214
Crawford, Margaret, 230
Cross, Brian, 136, 142
Cube, Ice, 128, 129, 130, 138, 139, 140, 141, 142, 145

Dahl, Robert, 67
Damon, George, 7
Davis, Mike, 12, 137, 139
Dear, Michael, 13 (n 2), 17-23
DeBord, Guy, 136
DeMille, Cecil B., 5
Denton, N., 44
Deukmejian, George, 156
Di Marco, Maureen, 198
Dog, Tim, 138
Dogg, Snoop Doggy, 128, 129, 130, 132, 135, 136, 144, 145
Dre, Dr., 128, 129, 130, 131, 132, 135, 136, 137, 142, 143, 144, 145
Dreiser, Theodore, 61
Duany, Andres, 256
Duke, Bill, 129
Dunne, John Gregory, 2-3
Dupont-Walker, Jacquelyn, 34
Dykstra, Clarence, 6

Subject Index

About the Contributors

ADOBE LA (Architects, Artists, and Designers Opening the Border Edge of Los Angeles) is a group of Latino architects and designers working on transborder issues in Los Angeles.

Ulises Diaz is a graduate of the Southern California Institute of Architecture (SCI-Arc) and member of ADOBE LA, where he is Project Manager of "Cultural Explainers," public art works in three culturally diverse communities in Los Angeles. In addition, he is Project Manager for Barrio Action Group, Inc., and Cofounder of the Inner City Murals Project, along with a group of young graffiti artists from downtown Los Angeles. He is on the SCI-Arc Design Faculty and adviser for SCIUDAD (Latin American Students of Architecture at SCI-Arc).

Gustavo Leclerc is a designer, artist, and founding member of ADOBE LA, where he is Project Manager of Laboratorio de Experimentación Urbana and *Bordergraphies/Huellas Fronterizas: Retranslating the Urban Text in Los Angeles and Tijuana.* He is a graduate of the Universidad Veracruzana, Mexico, in architecture and has extensive experience in designing architectural projects and in teaching at the Facultad de Arquitectura, Universidad Veracruzana. He is on the SCI-Arc Design Faculty and has lectured throughout the United States on architecture, design, art, and cultural criticism.

Alessandra Moctezuma is an artist and founding member of ADOBE LA. She is Project Manager of the Art for Rail Transit program of the Los Angeles MTA. In 1994, she received an M.F.A. from UCLA and is a recipient of various awards including the UCLA Arts Council Award and the Carlos Almaraz Scholarship. She is a member of Collage Ensemble, has participated in group exhibitions and performances, and has lectured widely on art, Latino culture, and vernacular architecture. She produced the videos *Andale, Andale; Frida;* and *Urban Revisions.* She is on the Design Faculty at SCI-Arc.

Leda Ramos is an artist and member of ADOBE LA, where she is Project Manager for the +Urbarte publication, *Los Angeles Work Sites and Women: Transformation of Public Space.* She is a College Art Association Fellow at the Getty Center for the History of Art and the Humanities (L.A. as Subject: The Transformative Culture of L.A. Communities) and a member of the Design Faculty at SCI-Arc. She is an M.F.A. graduate of Rutgers University, has participated in group and solo exhibitions, and has served as Multicultural Coordinator at the Los Angeles Municipal Art Gallery. She has a B.A. in sociology from the University of California, Santa Barbara, and a diploma from the Pennsylvania Academy of the Fine Arts.

Elpidio Rocha is an architectural theorist, designer, educator, and pioneer in the development of interdisciplinary art/design process. He teaches at the University of Oregon at the School of Architecture and Applied Arts and is a member of ADOBE LA. He was formerly on the faculty at the USC Urban Studies Program, School of Public Administration; Architecture Department, Cal State Polytechnic University Pomona; and Kansas City Art Institute. He has lectured and written extensively in the United States and Mexico on Chicano art, design, environment, and architecture.

Todd Boyd is Assistant Professor of Critical Studies at the University of Southern California School of Cinema-Television. His work has appeared in journals such as *Public Culture, Filmforum, Cineaste,* and *Wide Angle.* As a contributing columnist, he has written on popular culture for the *Chicago Tribune* and the *Los Angeles Times.* He is coeditor of *Out of Bounds: Sports, Media, and the Politics of Identity* and is the author of *Am I Black Enough for You? Popular Culture From the 'Hood and Beyond,* both forthcoming in 1996.

Leo Braudy is Professor of English at the University of Southern California.

Alida Brill is a social critic who writes about women, culture, and politics. She is the coauthor (with Herbert McClosky) of the award-winning book *Dimensions of Tolerance: What Americans Believe About Civil Liberties* (1983). She is the author of *Nobody's Business: The Paradoxes of Privacy* (1990). She edited the recently published book *A Rising Public Voice: Women in Politics Worldwide,* presented by the United Nations to each official country delegate at the U.N. Conference on the Status of Women held in Beijing in 1995. She serves on a variety of boards and professional committees and for several years was the program officer at the Russell Sage Foundation, where she directed a national program of research on gender and institutions. She publishes and lectures widely on the problems of power and gender, especially as they relate to sexual choice and behavior.

Margaret Crawford teaches architectural theory at the Southern California Institute of Architecture.

Michael Dear is Director of the Southern California Studies Center and Professor of Geography at the University of Southern California. His most recent publication is *Malign Neglect* (1993), a book about homelessness in Los Angeles (written with Jennifer Wolch). In 1995-1996, he was a Fellow at the Center for Advanced Study in the Behavioral Sciences at Stanford University.

Jacquelyn Dupont-Walker worked with Rebuild LA and is now with Ward Economic Development Corporation.

Gary A. Dymski is Associate Professor of Economics at the University of California, Riverside, and is Research Associate of the Economic Policy Institute in Washington, D.C. After completing his Ph.D. in economics (University of Massachusetts, 1987), he was an Assistant Professor of Economics at the University of Southern California and at the University of California, Riverside. He was selected as a Research Fellow in economic studies at the Brookings Institution in 1985-1986. He is on the editorial boards of the *International Review of Applied Economics* and *Geoforum* and previously served on the editorial board of the *Review of Radical Political Economics*. He has published numerous articles in academic journals and edited volumes on topics including the theory of bank behavior, the evolution of the U.S. financial system, race and credit, borrower-lender behavior in the LDC debt crisis, asymmetric information in the credit market, and the theory of exploitation. He is coeditor of *Transforming the US Financial System: An Equitable and Efficient Structure for the 21st Century* (with Gerald Epstein and Robert Pollin) and *New Directions in Monetary Macroeconomics: Essays in the Tradition of Hyman P. Minsky* (with Robert Pollin).

Robert Fishman is Professor of History at Rutgers University, Camden, New Jersey. He is the author of *Urban Utopias in the Twentieth Century* and *Bourgeois Utopias: The Rise and Fall of Suburbia*.

Robbert Flick is Professor of Fine Arts and Director of the Matrix Program for computers in the arts at the University of Southern California. A Los Angeles-based artist, he works with conceptually based photographic inventories of the changing landscapes, structures, and conditions in the city and surrounding areas. For the last 5 years, he has worked digitally, creating works that function simultaneously as art objects and as part of the USC database ISLA (Information Systems for Los Angeles). His work was most recently included in exhibitions at the Los Angeles County Museum of Art and Craig Krull Gallery at Bergamot Station, Santa Monica. Born in Holland, he did his undergraduate work at the University of British Columbia and received his M.F.A. from UCLA.

Stuart A. Gabriel is Professor of Finance and Business Economics in the Graduate School of Business Administration at the University of Southern California and a Visiting Scholar at the Federal Reserve Bank of San Francisco. He has published extensively on housing and mortgage markets and finance, urban and regional economics and policy, and population mobility. He serves on the editorial boards of *Real Estate Economics* and the *Journal of Real Estate Finance and Economics.* He is also a member of the Advisory Board of the Fannie Mae Office of Housing Policy Research and is a Director of the American Real Estate and Urban Economics Association, the Los Angeles Economic Roundtable, and the Los Angeles Community Design Center. He has recently consulted for the U.S. Department of Housing and Urban Development, Fannie Mae, the American Bankers Association, and E&Y Kenneth Leventhal. Prior to joining the USC faculty, he served on the economics staff of the Federal Reserve Board. He holds A.B., M.A., and Ph.D. degrees in economics from the University of California, Berkeley.

Genevieve Giuliano is Associate Professor and Director of the Lusk Center Research Institute in the School of Urban Planning and Development, University of Southern California. She received her Ph.D. in social science from the University of California, Irvine. Her research interests include transportation policy evaluation, land use and transportation relationships, and travel behavior. She is coeditor of the international journal *Urban Studies,* a member of the editorial board of *Transportation Research,* and a Research Fellow at the Lincoln Institute of Land Policy.

Harlan Hahn is Professor of Political Science at the University of Southern California. His book on the violence of the 1960s, *Ghetto Revolts* (coauthored with Joe R. Feagin), was nominated for the Pulitzer Prize. Although he maintains a strong interest in urban politics, much of his current work has focused on gaining equal rights for people with disabilities.

Eloise Klein Healy is a Los Angeles-based poet.

Guilbert C. Hentschke is Dean of the School of Education and Professor of Educational Policy and Administration at the University of Southern California. His most recent work includes education policy, school governance, higher education administration and management, urban teacher education reform, and school performance of youth from low-income families. He recently completed (with Brent Davies) "School Autonomy: Myth or Reality: Developing an Analytical Taxonomy" in *Educational Management and Administration.* He currently serves on the boards of the National Center for Education, the Galaxy Institute for Education, Southwest Regional Laboratory (SWRL), and Policy Analysis for California Education (PACE) and on the executive committee of Los Angeles Annenberg Metropolitan Project (LAAMP).

Greg Hise is an urban historian and Assistant Professor in the School of Urban and Regional Planning at the University of Southern California. He is the author of *Magnetic Los Angeles: Planning a Postwar Metropolis* (forthcoming). Currently, he is investigating the formation of industrial districts for aircraft and film production and how these new technologies inform perceptions of cities and regions.

Paul C. Hudson is President and Chief Executive Officer of Broadway Federal Savings and Loan.

Lillian Kawasaki works on environmental issues for the City of Los Angeles Environmental Affairs Department.

Mark Kroeker is Deputy Chief of the Los Angeles Police Department.

Abby J. Leibman is Executive Director of the California Women's Law Center in Los Angeles.

William Loos is Medical Director of the Los Angeles County Department of Health Services.

Rubén Martínez is a critic, poet, and educator. He is Regional Head of the Pacific News Service and regular cohost of PBS TV's *Life and Times.*

The Reverend Cecil Murray is Pastor of the First AME Church in Los Angeles.

Edward J. W. Park is Assistant Professor in the Department of Sociology and the Program in American Studies and Ethnicity at the University of Southern California. He received his Ph.D. from the Graduate Group in Ethnic Studies and a master's degree in city planning, both at the University of California, Berkeley. His research topics include the impact of contemporary regional economic restructuring on urban race relations and the Los Angeles racial politics since the Civil Unrest of 1992.

Laura Pulido is Assistant Professor of Geography at the University of Southern California, where she teaches courses on environmentalism and racial issues. She is the author of *Environmentalism and Economic Justice: Two Chicano Struggles in the Southwest.* She is active in local environmental affairs and is a member of the Labor/Community Strategy Center.

Steven B. Sample is President of the University of Southern California.

H. Eric Schockman is Associate Director of the Center for Multiethnic and Transnational Studies and is a faculty member of the Department of Political Science at the University of Southern California. He has served for more than a decade as a top administrator and consultant to the California State Assembly and to the City Council of Los Angeles. He is a recognized expert in public policy formation and governmental politics. His most recent work (completed with George Totten) is *Community in Crisis: The Korean-American Community After the Los Angeles Civil Unrest of April, 1992.*

Robert E. Tranquada is the Norman Topping/National Medical Enterprises Professor of Medicine and Public Policy at the University of Southern California. A graduate of Pomona College and Stanford University School of Medicine, he trained in internal medicine. He has held medical faculty positions since 1959, which have included deanships of schools of medicine at the University of Massachusetts and the University of Southern California. He is a member of the Institute of Medicine of the National Academy of Science.

John M. Veitch is Associate Professor of Economics at the University of San Francisco. After completing his Ph.D. in economics at Northwestern University, he was Assistant Professor of Economics at the University of Southern California from 1985 to 1992. While in graduate school, he did collaborative research with the noted macroeconomist Robert J. Gordon; together, he and Gordon published a widely cited paper on investment in the U.S. economy. He has won teaching awards and attracted substantial grant support for his innovative work on using computer technology in the economics classroom. In addition to his research with Gary Dymski on finance, race, and development in Los Angeles, he has published numerous academic articles on topics including economic history, econometric theory, and macroeconometrics.